The information for your exponential success

Infonential, Inc.

Gateway to Nanotechnology

An Introduction to Nanotechnology for Beginner Students and Professionals

Paul Sanghera, Ph.D.

Gateway to Nanotechnology: An Introduction to Nanotechnology for Beginner Students and professionals

Published by

Infonential, Inc.

A California Corporation.

http://www.infonentialinc.com

email: info@infonentialinc.com

ISBN-10:1-4392-3087-0

ISBN-13: 978-1-4392308-7-9

This publication is designed to provide accurate and authoritative information on the covered subject matter. It can be used by students and professionals in nanotechnology. However, it is sold with the understanding that the publisher is not engaged in offering technical, legal, or other professional service. If such assistance is required, the service of a competent professional should be sought.

Warning and Disclaimer

The information published in this book has been obtained by Infonential, Inc. from sources believed to be reliable. We have made our best effort to make this material as comprehensive within the scope of this book and as accurate as possible. However, because of possible errors such as human, mechanical, or technical, the publisher and the author does not guarantee the accuracy, adequacy, or completeness of any information and are not responsible for any errors or omissions and the results obtained from such information

To

The great scientist Richard Feynman

Who

Inspired me

To

Expand my research horizons

Beyond physics

And

To my son Adam Sanghera

For

Giving me the opportunity

To

Enjoy watching him to grow

From a nanoman to a macroman

Table of Contents

The Gateway Book Series

The purpose of *a Gateway book* is to introduce the reader to a huge, complex field in an easy-to-understand way. The goal is to present a wide spectrum of topics on the subject in a cohesive, concise, yet comprehensive fashion. Once introduced to the breadth of the subject in this fashion, the reader can choose to explore the depth of a topic of interest.

The lack of information is the problem of the past ages. The problem of the current age, the information age, is that we have too much information and too little time to absorb it. We are being bombarded with all kinds of needed and (mostly) un-needed information from all sides. This situation has created the need for information products that contain well defined, precise, to the point, quick, and "just what the customer needs" information. The *Gateway Book Series* is a response to this market need. In other words, a *Gateway book* is based on the philosophy that in this fast paced information age, nobody has the time to prowl through web pages or struggle with a 900 pages long book to find just that little piece of needed information.

Bottom line: A *Gateway book* offers maximum learning in minimum time. No fluff, just the nuggets of the needed information presented in an easy to understand format.

About the Author

Dr. Paul Sanghera, an educator, scientist, technologist, and an entrepreneur, has a diverse background in major fields on which nanoscience and nanotechnology is based including physics, chemistry, biology, computer science, and math. He holds a Master degree in Computer Science from Cornell University, a Ph.D. in Physics from Carleton University, and a B.Sc. with triple major: physics, chemistry, and math. He has taught science and technology courses all across the world including San Jose State University and Brooks College. He has authored and co-authored more than 100 research papers published in well reputed European and American research journals. At the world class laboratories such as CERN in Geneva, Switzerland, and Nuclear Lab at Cornell, he has participated in designing and conducting experiments to measure time in nanoseconds and size in nanometers to study the basic building blocks of matter which compose nano particles.

As a technology manager, Dr. Sanghera has been at the ground floor of several technology startups. He is the author of several best selling books in the fields of science, technology, and project management. He lives in Silicon Valley, California.

About This Book

Format: lecture notes which are:

❖ self-contained

❖ presented in a logical and easy-to-follow sequence

❖ Comprehensive and complete within the scope of this book

Who can benefit from this book?

❖ Professionals who want to join the field of nanotechnology.

❖ Beginner Students of nanotechnology.

❖ Managers and executives for a quick introduction to nanotechnology.

1

Introduction to Nanotechnology

Learning Objectives

1. Definitions

2. Scientific Background

3. Material science

4. A Brief History of Nanotechnology

What is Nano?

✓ **Nano** means very small.

✓ Precisely speaking ➜ **Nano** means billionth, means 1/1,000,000,000

✓ **Nano** can also be written as 10^{-9}

✓ One **nanometer**: Billionth part of a meter

✓ One **nanosecond**: Billionth part of a second

✓ One **nanometer** is about 80,000 times smaller than the thickness (diameter) of a human hair.

✓ **Nanoscale**. The scale that deals with nano sizes.

 *What is 10^{-9}? Can you tell me more about powers of ten, please?*

Powers of Ten

Helpful in handling very big and very small numbers that the scientists deal with

Table 1.1 Metric system of numbers in powers of ten used to describe measurements such as size.

Metric prefix	Symbol	Name	Number	Powers of Ten
Exa	E	Quintillion	1,000,000,000,000,000,000	10^{18}
Peta	P	Quadrillion	1,000,000,000,000,000	10^{15}
Tera	T	Trillion	1,000,000,000,000	10^{12}
Giga	G	Billion	1,000,000,000	10^{9}
mega	M	Million	1,000,000	10^{6}
kilo	K	Thousand	1,000	10^{3}
unity	---	One	1	10^{0}
milli	m	Thousandth	1/1,000	10^{-3}
micro	μ	Millionth	1/1,000,000	10^{-6}
nano	n	Billionth	1/1,000,000,000	10^{-9}
pico	p	Trillionth	1/1,000,000,000,000	10^{-12}
femto	f	quadrillionth	1/1,000,000,000,000,000	10^{-15}
atto	a	quintillionth	1/1,000,000,000,000,000,000	10^{-18}

Examples of usage

$5.00 \times 10^{9} m = 5.00$ billion meter

$5.00 \times 10^{-9} g = 5.00$ nanogram

$5.00 \times 10^{-3} g = 5.00$ milligram

$5.00 \times 10^{-9} m = 5.00$ nanometer

What is Nanoscience and Nanotechnology?

 **Nanoscience** is the study of matter and material objects such as molecules and structures with at least one dimension in the range of 1 nm to about 100 nm.

 **Nanotechnology** is the technology that deals with the manipulation of materials on an atomic or molecular scale measured in billionths of a meter, nanometers. It's an application of nanoscience.

Relationship between nanoscience and nanotechnology

Same as relationship between any science and the corresponding technology

Dialectical

Nanotechnology uses the knowledge of nanoscience, and in turn helps nanoscience to advance, for example, by providing nano tools.

Why Nanotechnology?
What's the big deal?

♦ **Science need**. Need to understand substances at molecular & atomic levels in order to design, build, and manipulate substances at macro-levels.

♦ **Technology need**. Need to acquire the ability to make extremely small things. Limitations largely come from the technology and not from the underlying physical laws.

♦ **Applications**. At a nanoscale, many substances may possess unique characteristics because the material world at this scale is governed by different physical laws (quantum physics as opposed to classical physics). This may give rise to new products and capabilities.

♦ **Life need**. Miniaturization impacts the sustainability of the *triangle of life*:

- Design and manufacturing of materials and tools, which include products and machinery

- Storage and processing of information

- Storage and transport of energy

 *What is the triangle of life?*

Triangle of Life

The triangle life is composed of three interdependent elements that support life on earth: materials and tools, information, and energy.

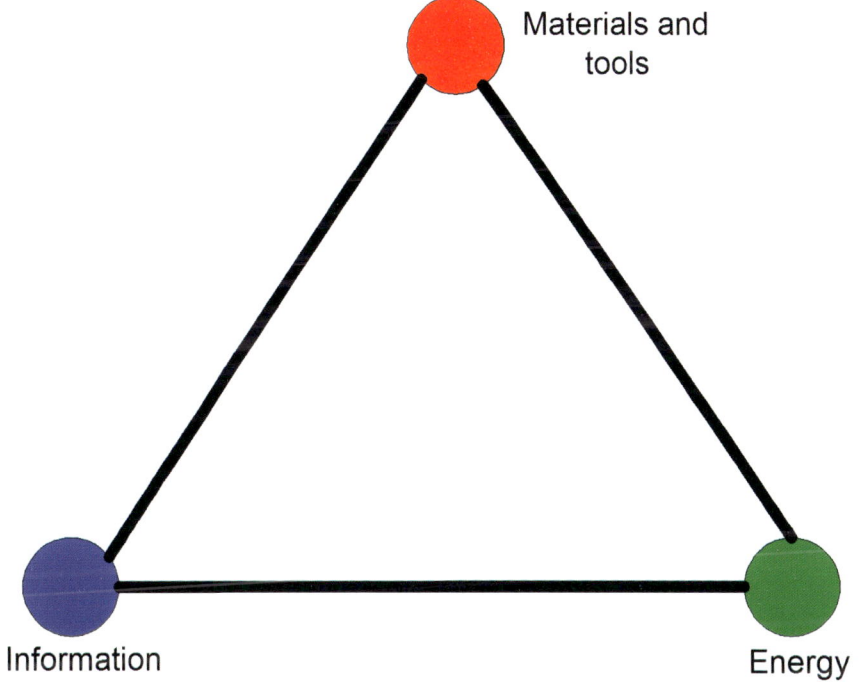

Figure 1.1 Triangle of life impacted by nanotechnology

Yes, we need nanotechnology, but even more than that:

Nanotechnology is the result of a natural progression of several disciplines to a common unification point ➔

Nanotechnology: A Grand Unification

Nanoscience, the underlying science of nanotechnology, is based on unification of a multitude of sciences at the junction of nanoscale: physics, chemistry, biology, computer science, engineering, material science…

 **Definitions:**

Physics; the mother of all sciences. Science that deals with understanding the universe and the systems in the universe in terms of fundamental constituents of matter (such as atoms, electrons, and quarks) and the interactions between those constituents.

Chemistry. Study of matter and energy as related to physical and chemical changes.

Biology. Scientific study of living systems such as organisms and cells.

Computer science. Study of the fundamentals of information and computation and their implementation and applications in computer systems.

Engineering. The discipline that deals with acquiring and applying scientific and technical knowledge to design, and analyze and construct devices and systems for practical purposes.

Material science. Interdisciplinary field based on investigating the relationships between the structures of materials and the properties of materials, and applying this knowledge to various areas of science and engineering.

Yes, we also need math, the language of science.

Nanotechnology has brought *material science* to the forefront.

 *Hold on, can you please tell me more about material science?*

Fundamentals of Material Science

➢ **Common sense.** Materials are important in the material world. Think of the ages in human history defined by materials: *Stone Age, Bronze Age, Steel Age...*

➢ **Central Theme.** Understand the existing material by exploring relationships among different material aspects. The core relationships studied in material science are the relationships of the properties of the material to its structure ➔ Use this understanding to manipulate existing materials and create new materials.

➢ **Major determinants of the structure of a material**:

    ♦ The constituent chemical elements that compose it

    ♦ The way the material was processed into its existing (final) form

These combined with the underlying laws of thermodynamics constitute the microstructure and therefore the properties of the material.

 *By the way, what's thermodynamics?*

What is thermodynamics?

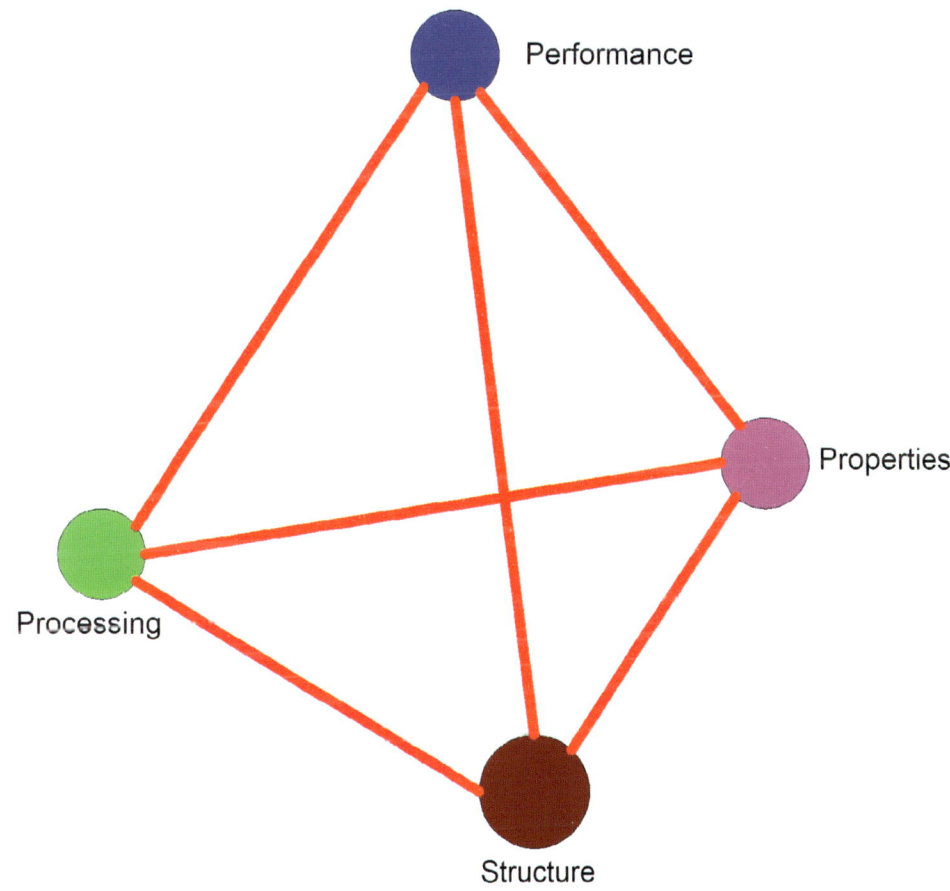

Thermodynamics. The branch of physics that studies the dynamics of heat, that is, the dynamics of energy. To be precise, it studies the effects of changes in temperature, pressure, and volume on physical systems at the macroscopic scale.

Back to Material Science:

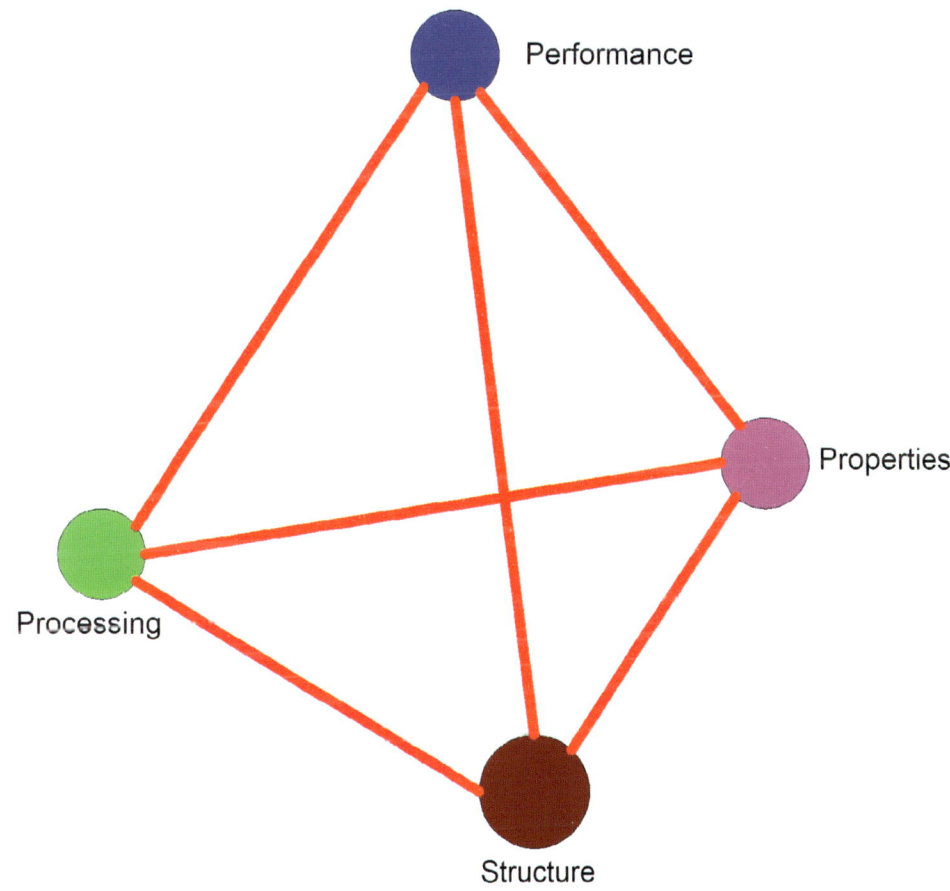

Figure 1.2 Relationship among performance, properties, processing, and structure.

History of Science

A Journey from Macro to Nano

Fact of Nature: Things expose their similarities more at microcale and look more different at macro scale ➜ There is unity at micro scale behind diversity at macro scale.

We, human are macro beings

We started studying different things in different fields at macro scale by using macro tools ➜ Different disciplines of science

The history of the development of science is the journey of scientific studies from macro scale to micro scale, for example, from organisms to cells in biology and from planets to atoms and even deeper, the subatomic particles in physics.

The broader definitiion of micro includes nano

At micro scale things start looking more similar or unified. **Examples:**

✓ All diverse forms of life are made of the same building blocks called cells, and all cells are made of the same yet smaller building blocks called biomoleculs.

✓ All different macro materials are made of a few basic building blocks called atoms, and all atoms have the same basic structure and obey the same physical pronciples.

Ingredients of Unification

At nanoscale, studying different things in different fields:

- ◆ Involves the understanding of the same science: quantum mechanics. In other words, principles of quantum mechanics, also called quantum physics, govern the behavior of material things at nanoscale.

- ◆ Requires the same kind of techniques and manipulation tools: the nano tools such as electron microscopes.

- ◆ Faces the same or similar research problems and issues

Example of Mice and Men

Looking at macro scale:

Man and mouse are obviously different from each other.

Looking at nanoscale:

The genomes (set of genes) of mice and men are 99% similar

The point here is:

A while ago during its journey from macro to nano, physics discovered the basic law of nature: unity behind diversity on which the theories of unification are based. Unification brought by nanotechnology is just another expression of that law. The expression or display of unification depends on the scale.

History of Nanotechnology

On One Page

December 29, 1959, APS Meeting at Caltech.
Physicist Richard Feynman gives a talk*: There is plenty
of room at the bottom. This is now considered as the
beginning of the nanotechnology field.

1974. Norio Taniguchi first defines the term
nanotechnology to refer to "production technology to
get the extra high accuracy and ultra fine dimensions".

1981. Scanning Electron Microscope invented.

**Richard Feynman
(1918-1988)**

1985. Buckyball, a carbon structure of intense interest to
nanotechnology, discovered.

1986. K. Eric Dexler. Further exploration of the nanotechnology
concept and possibilities in the book: *Engines of Creation: The Coming
Era of Nanotechnology*

1986. Atomic Force Microscope, another nano tool invented.

2000. President Bill Clinton announces the US National
Nanotechnology Initiative:

> *http://www.nni.gov*

Recall, it's all about the scale ➜

Scale of Things

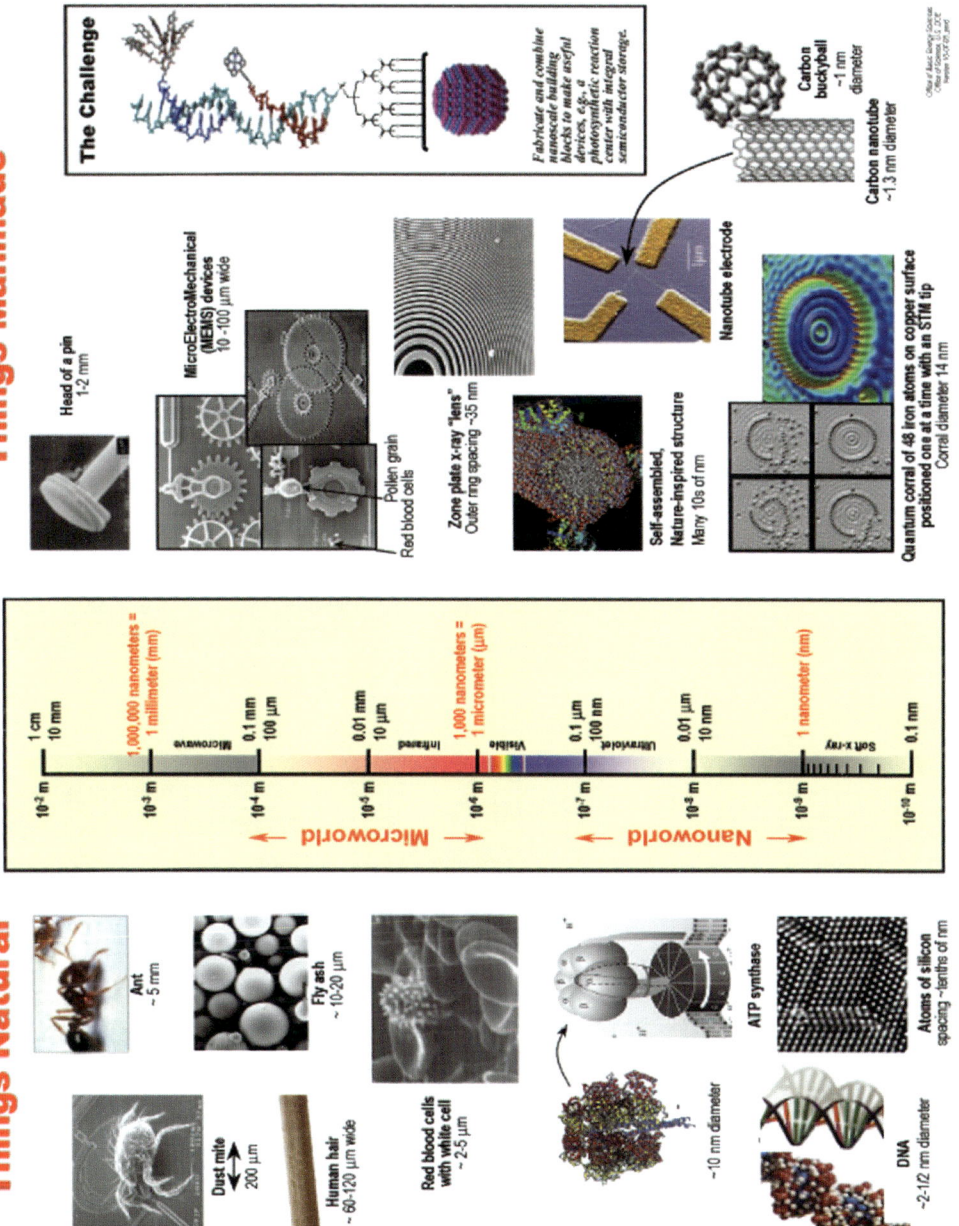

Figure 1.3 Scale of things.

Nature Works at Nanoscale

Nature performs its most important work at nanoscale which exhibits itself both at macro and nano scales:

▪ The properties of materials at macro scale (called bulk materials) are often related to their structure designed at micro and nano scale.

▪ Some materials have nanostructures exposed to the macro world in a more pronounced and direct way. Tiny bumps on the surface of a moth's eye are an example.

▪ These patterns are smaller than the wavelength of visible light ==>

    o Moth's eye can absorb more light.

    o Moth can see much better than human in dim or dark conditions.

Figure 1.4 The nano-sized hexagonal shaped patterns on the surface of a moth's eye.

This is interesting; can you please give me another example from the nature?

Nature Works at Nanoscale: Another Example

Multilayer nanoscale patterns on the surface of a butterfly's wings:

Figure 1.5 Wings of the male *morpho rhetenor* appear bright blue. **Why?**

The pattern size falls in the range of the visible light wavelengths

Due to multiple layers:

◆ Constructive interference among the reflected light waves around 450 nm wavelength, which is the wavelength that corresponds to the blue color.

◆ Destructive interference at other wavelengths.

Wings appear bright blue.

Nature Works at Nanoscale: So What?

Using nanotechnology:

♦ Study *nature's* works at nanoscale.

♦ Use the knowledge from the study to:

✓ Manipulate existing materials

✓ Build new materials

Example:

DNA

But nature is beyond limits…so nanotechnology must be a very broad field. Can you categorize it?

 Yes, nanotechnology embraces both living and non-living systems and structures:

Two Types of nanotechnology:

♦ Wet nanotechnology about living things

♦ Dry nanotechnology about non-living material

Two Types of Nanotechnology

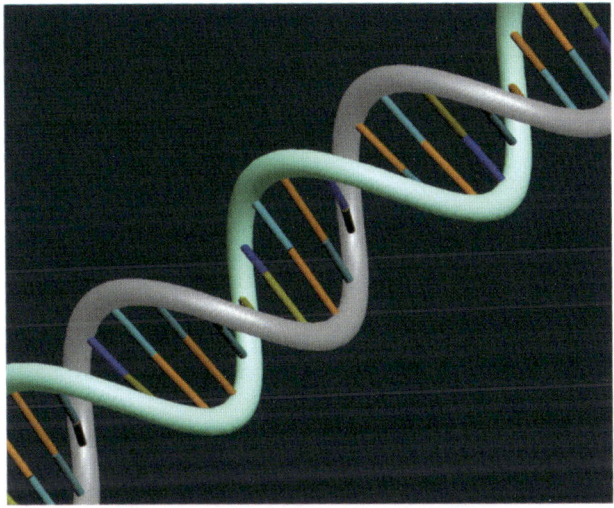

Wet Nanotechnology:

- All the nano-machinery of cellular life (and viruses)

- Biotechnology is a form of Nanotechnology

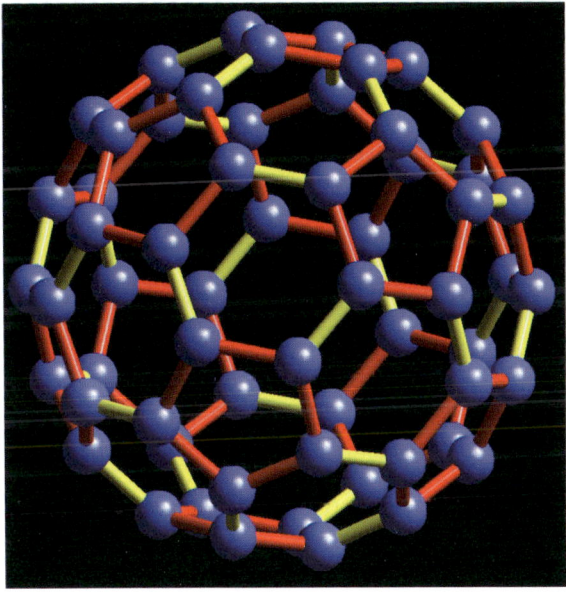

Dry Nanotechnology:

- Electrical & thermal conduction

- Great strength, toughness, and high temperature resistance

NEMS are examples of dry nanotechnology ➜

Nanoelectromechanical systems

First, there were MEMS

- **Microelectromechanical systems (MEMS).** Integration of mechanical structures (e.g. moving parts) with micro electronics.

  o Also referred to as micromachines (in Japan) and micros system technology (in Europe).

  o Generally range in size from a millimeter to a micrometer.

- **Nanoelectromechanical Systems (NEMS).** The nano versions of microelectromechanical systems (MEMS), that is, nanoscopic system that has characteristic length of less than 100 nm, and combines the electrical and mechanical components.

Both MEMS and NEMS exploit large surface area to volume ratio

Surface effects such as electrostatics dominate the volume effects such as inertia.

Summary and Conclusions

➢ Significant fundamental understanding of nanoscience and nanotechnology has been achieved over the last few decades.

➢ Nanoscience is the study of matter and material objects such as molecules and structures with at least one dimension in the range of 1 nm to about 100 nm.

➢ Nanotechnology is the technology that deals with the manipulation of materials on an atomic or molecular scale measured in billionths of a meter (nanometers). It's an application of nanoscience.

➢ Nanotechnology is the resulting platform of several scientific disciplines and provides the capability to develop advanced solutions to great scientific and technical challenges.

➢ Nature does its most important design work at nanoscale from which nanotechnologists can learn by studying some natural structures and processes.

➢ Nanotechnology can be broadly categorized into two types: *wet* (e.g. nanobiotechnology), and *dry* (based on non-bio materials).

Self Test Exercises

1   **True or False**: Nanoscale is larger than micro scale but smaller than macro scale.

2   Write the following quantities in powers of 10:

   A. 75 nanometer

   B. 500 nanogram

   C. 35 millimeter

3   Convert the following quantities:

   A. 0.25 g = _____ µg

   B. 3.5 µL = _____ nL

   C. 560 mm = _____ cm

   D. 2.5 x 10^8 µg   = _____ kg

E. 0.055 m = _____ mm

F. 136 mL = _____ L

D. 35 millimeter

4 Name some fields that merge at the nano junction .

5 Give examples of two structures that are studied in wet nanotechnology.

6 Give examples of two structures that are studied in dry nanotechnology.

7 Name two nano tools.

8 The material world at nanoscale is governed by what branch of physics?

Write Your Own Notes

2

Business of Nanotechnology

Learning Objectives

1. *Funding as an Indicator*

2. *Government Funding*

3. *Venture Capital*

4. *Careers in Nanotechnology*

5. *Jobs and positions in nanotechnology*

6. *Summary and Conclusions*

7. *Self Test Exercises*

Write Your Own Notes

Government Funding

Why is Funding Important?

❖ Government funding is an important indicator of the current interest in the field from:

- ◆ Government

- ◆ Private investors

- ◆ Other parties

❖ Initial funding gives some credibility to a new emerging field or the new topics in the existing field to attract private investors.

❖ Government funding builds the much needed momentum in a field.

❖ Government funding helps advance the field by making new discoveries and inventing new techniques and methods.

❖ Private investors are usually not interested in the research that generates no immediate profit but that is important to bring the field to a point where it could give rise to useful applications.

Can you please give me an example of funding in nanotechnology?

The Rising NNI Annual Budget

Note the linear rise in the NNI annual funding:

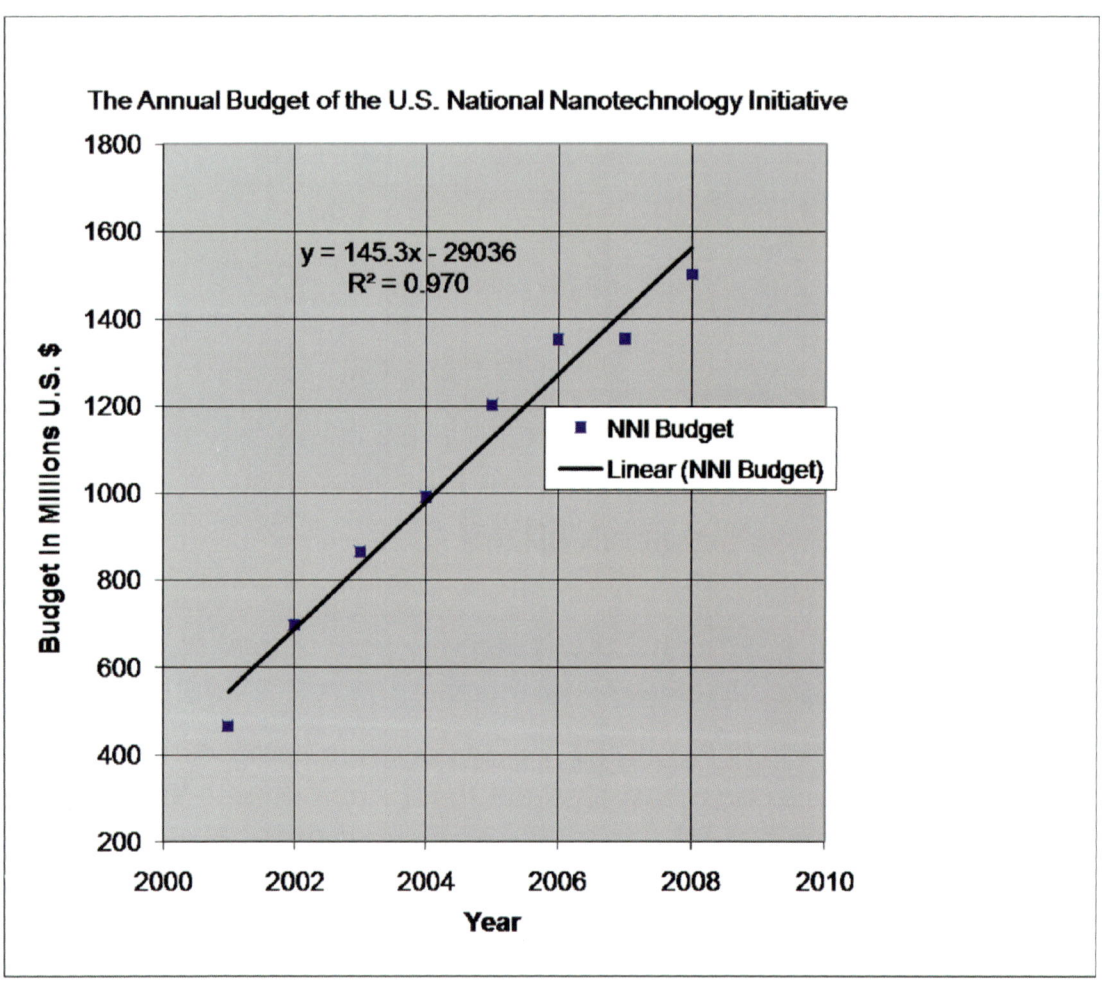

The Annual Budget of the U.S. National Nanotechnology Initiative

$y = 145.3x - 29036$
$R^2 = 0.970$

Distribution of Funding

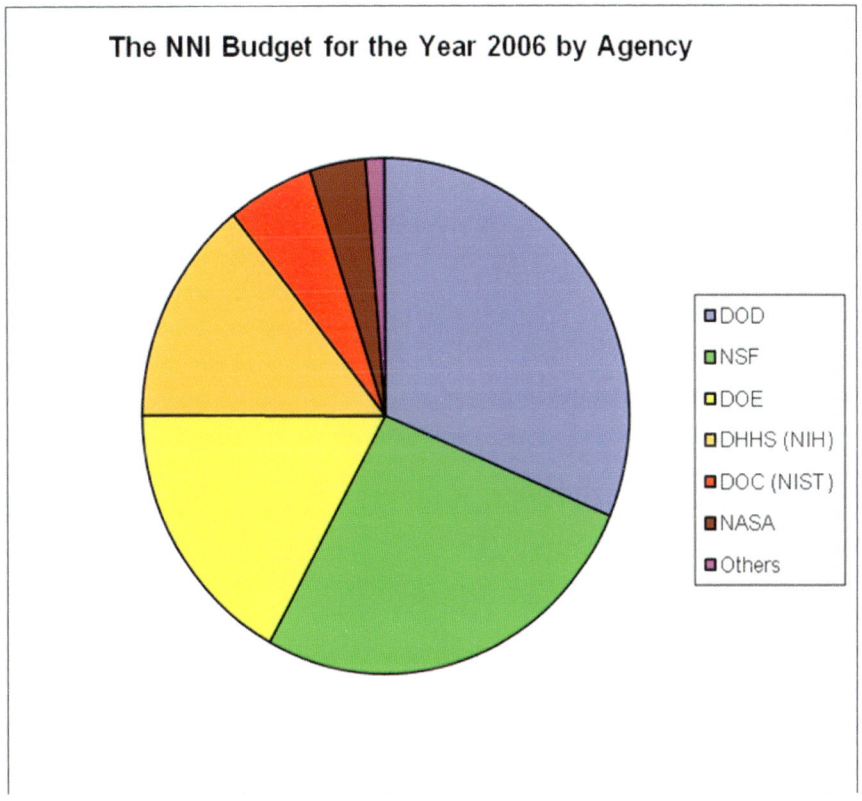

The NNI Budget for the Year 2006 by Agency

Legend:
- DOD
- NSF
- DOE
- DHHS (NIH)
- DOC (NIST)
- NASA
- Others

BTW, what is NNI?

NNI and Government Agencies

 **National Nanotechnology Initiative (NNI):** A US Government program to unify the nanotechnology efforts by the government at one platform.

NNI uses its budget to support a broad range of nanotechnology programs among various U.S. government departments and agencies. The list of these agencies includes the following:

- **Department of Commerce (DOC), and National Institute of Standards and Technology (NIST):** http://www.commerce.gov and http://www.nist.gov

- **Department of Defense (DOD)**: www.nanosra.nrl.navy.mil

- **Department of Energy (DOE):** www.sc.doe.gov

- **Department of Homeland Security (DHS):** www.dhs.gov

- **Department of Health and Human Services (DHHS):** http://www.hhs.gov

- **Department of Justice (DOJ)**: www.usdoj.gov

- **Environmental Protection Agency (EPA):** www.es.epa.gov

- **National Institute of Health**: www.nih.gov

- **National Institute of Standards and Technology (NIST)**: www.nist.gov

- **National Aeronautics and Space Administration (NASA)**: www.ipt.arc.nasa.gov

- **National Science Foundation**: www.nsf.gov

- **U.S. Department of Agriculture (USDA)**: www.usda.gov

Nanotech Worldwide Government Funding

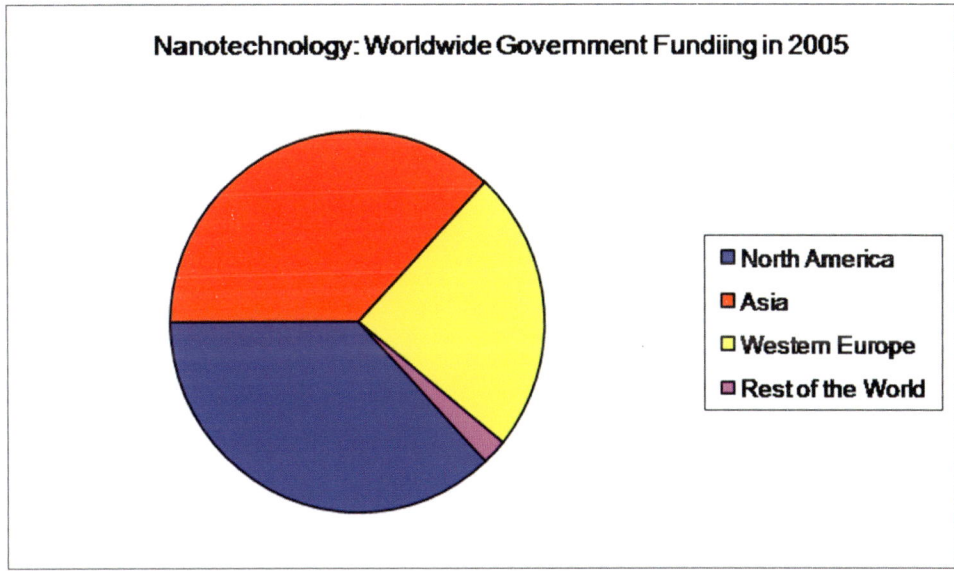

Nanotechnology: Worldwide Government Fundiing in 2005

Region	Budget in Billions of US $
North America	1.7
Asia	1.7
Western Europe	1.1
Rest of the worls	0.1

Note that Asia is putting as much government funding into nanotechnolofy as Northe America.

Nanotech Venture Funding

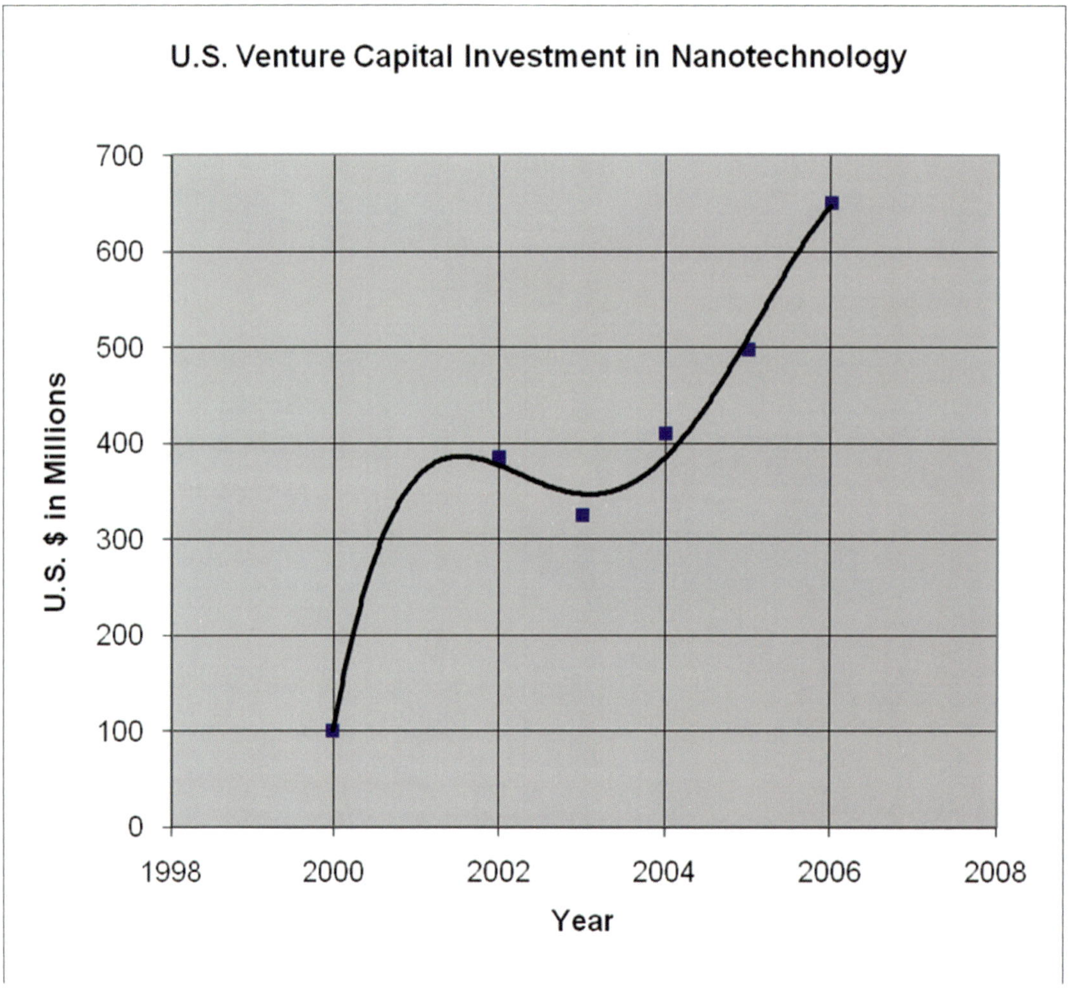

U.S. Venture Capital Investment in Nanotechnology

 Why all this funding? Will nanotechnology create any jobs?

Nanotechnology Workforce

Estimated need: Approximately 2 million workers worldwide by 2015
(**REF: National Nanotechnology Initiative**):

- **USA:** 0.8-0.9 million
- Japan: 0.5-0.6 million
- **Europe**: 0.3-0.4 million
- **Asia Pacific** (excluding Japan): 0.2 million
- **Other regions**: 0.1 million

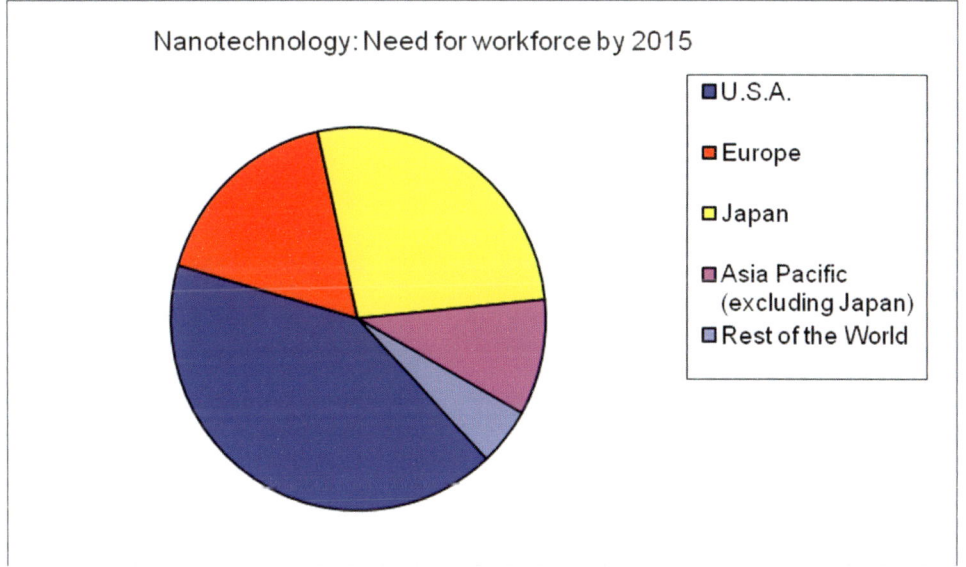

 *What are some of the areas in which nanotechnology jobs will be created?*

Nanotech Career Areas

Current applications of nanoscience and nanotechnology, and therefore the career opportunities, exist in a multitude of areas including the following:

- ✓ Physics
- ✓ Chemistry
- ✓ Engineering
- ✓ Biology, biotechnology, and bioinformatics
- ✓ Electronics and semiconductor industry
- ✓ Environmental monitoring and control
- ✓ Food science including quality control and packaging
- ✓ Material science including textiles, polymers, and packaging
- ✓ Medical fields
- ✓ Optoelectronics
- ✓ Pharmaceuticals including drug delivery and cosmetics
- ✓ Sports equipment
- ✓ University and federal lab research
- ✓ Corporate research
- ✓ And many more

 *Will nanotech jobs be in easily identified nanotech companies?*

The Scattered Nanotech Jobs

In addition to clearly identified nanotech companies, the nanotech jobs may also be spread over several fields affected by nanotech. So, pay attention to the developments in the following fields as well:

- Agricultural Science
- Auto and aerospace industries
- Business
- Environmental Science
- Ethics
- Forensic Science (the word *forensic* means legal).
- Law
- Medicine
- Military
- National security

 *Can you please actually name some current positions?*

Some Actual Names for Nanotech Positions

- ✓ Research assistant
- ✓ Lab technician
- ✓ Electron microscope expert
- ✓ Material synthesis technician
- ✓ Scientific project manager
- ✓ Scientific program manager
- ✓ Engineer
- ✓ Engineering manager
- ✓ Trainer
- ✓ Technical writer
- ✓ Nanotechnologist
- ✓ Scientist

Summary and Conclusions

♦ Governments all over the world has poured money equivalent to billions of dollars into the nanotechnology research and development. This funding has boosted the advancement of the nanotechnology field.

♦ Venture capitalists' interest in nanotechnology supported by funding has also been on the rise.

♦ Due to its colossal scope, nanotechnology offers opportunities in an amazingly broad spectrum of areas and fields ranging from physics to sports.

♦ Nanotechnology is expected to generate 2 million jobs worldwide by year 2015; about 0.9 million in the U.S. alone.

♦ Some actual nanotech positions are already available at various companies and institutes.

Self Test Exercises

1 Explore nanotech positions at the following websites:

http://www.workingin-nanotechnology.com/

http://www.tinytechjobs.com

2 Asia is matching dollar with dollar on government funding in nanotechnology. What does it tell you about nanotechnology?

3 What is NNI?

4 What is NIH?

3

Applications of Nanotechnology

Learning Objectives

1. *Nanotech Products Space*

2. *Nanotechnology Enhanced Products*

3. *Biotech Products and Applications*

4. *Carbon Nanotubes*

5. *Biosensors and Chemical Sensors*

6. Carbon Nanotubes

7. Alternative Energy Applications

8. Learning from Nature

9. Summary and Conclusions

10. Self Test Exercises

Nanotechnology Product Space

Very broad product space:

❖ **Spatial dimension.** Products of all sizes:

- ✓ Nano
- ✓ Micro
- ✓ Macro

❖ **Temporal (time) dimension.** Both old and new products:

- ▪ **Old** products with enhanced properties and characteristics
- ▪ **New** products which did not exist before

❖ **Area dimension**. Products in a multitude of different areas.

 *What are the areas of nanotech products?*

Nanotechnology Applications and Products

Some Areas:

♦ Aerospace

♦ Automotive

♦ Biotechnology

♦ Chemical industry

♦ Cleantech (alternative energy)

♦ Consumer products industry

♦ Defense and security

♦ Electronics and semiconductors

♦ Textile

 Some nanotechnology-enhanced products are already available in many of these areas.

 *What are the nanotechnology-enhanced products?*

Nanotechnology Enhanced products

The first generation of nanotech products: enhance/improve the properties of existing products by using nanotechnology.

The consumer world is already exploded with the nanotechnology-enhanced products including the following:

- Bowling balls that are harder
- Golf balls that fly straighter
- Nano car wax that gives you a shinier looking vehicle because it fills in tiny cracks more effectively than the standard wax
- Smart clothing:
    - ✓ Pants that repel wax
    - ✓ Shirts that don't stain
    - ✓ Socks that don't stink due to the inclusion of nano-sized silver particles
- Tennis balls that last longer
- Tennis rackets that are stronger
- Evolution of video games from arcade size of the past such as Pong, Frogger, and PacMan to the modern more sophisticated games being played in homes: Games on platforms such as the X Box, Playstation, and Game Cube.
- Other electronic applications: Cell phones, flash drives, DVDs, digital cameras and displays, and film batteries.

All these enhanced products are possible due to the ability to work effectively and efficiently at the nanoscale. For example ➜

Nanotechnology Applications
Smart Clothing

How does it work?

- The clothing industry uses nanotech to make stain-repellent fabrics.

But how?

- A chemical process used during manufacturing forces liquids to bead up when spilled on a garment for easy wiping away.

- Socks that are made with nano silver particles give anti-microbial protection, that is, preventing bacteria and fungus that cause itchiness and smells.

Nanotechnology Applications:
Niny Bits

- The cosmetics industry already puts nano-particles in lotions, creams, and shampoos.

  Why?

- Nano-sized zinc oxide particles are used in sun creams.

  Why?

- These particles perform tasks such as absorbing ultra-violet rays; make the lotion *transparent* and *smooth* instead of sticky and white, and so on.

 *Any example from biotechnology?*

Nanotech Products in Biotechnology

Some examples:

♦ Bandages embedded with silver nanoparticles

♦ Drug delivery patches

♦ Man made skin: nanofabricated network

♦ Insulin biocapsules (under development)

♦ Buckyballs to deliver drugs.

Insulin Biocapsules:

♦ Made of a micro-fabricated silicon membrane that has nanoscale pores and is bound to a polymer well.

♦ It has a potential as an implantable system for steady and continuous insulin production and delivery into the bloodstream.

Insulin:

A polypeptide hormone that regulates the amount of sugar in the blood by stimulating cells; used to treat diabetes.

You mentioned drug delivery patches. Can you give me an example in this area please?

Transdermal Drug Delivery

 **Transdermal drug delivery**: A drug delivery method that uses the transdermal patch technology:

- Transdermal patch technology: approved by the U.S. FDA (Food and Drug Administration) in 1979.
- The U.S. market for transdermal patches has grown to $3 billion per year.
- Advantage:
  - Likely to improve the bioavailability of drugs because it avoids passing through liver metabolism first
  - Maintains more uniform drug plasma levels than that in injections

Oh, yes, and biosensors →

Biointerface

Biosensor Applications

The probe molecules could be:

✓ Proteins (e.g. enzymes and antibodies)

✓ Transition metal complexes (hemes)

✓ Nucleic acids

 *How about chemical sensors?*

Nanotechnology Applications
Chemical Sensors

Example: Use of a nanotube in sensing

♦ Every atom in a single-walled nanotube (SWNT) is on the surface
➔ exposed to the environment.

♦ Charge transfer or small changes in the charge-environment of a
nanotube can cause drastic changes to its electrical properties

♦ The change in the electrical properties of the nanotube is recorded
and therefore the change in the environment is sensed.

 *OK, it all makes sense, I guess. But, what are these carbon
nanotubes?*

Nanotechnology Applications
Carbon Nanotubes

♦ Carbon nanotubes are sheets of graphite (carbon) that are rolled up on themselves.

♦ Just a few nanometres across:

✓ Single walled nanotube (SWNT) diameter: 0.5-2.0 nm

✓ Multi walled nanotube (MWNT) diameter: 10-50 nm

♦ These ultra-strong cylinder-like nano objects can be put to enormous number of applications, for example:

o These tubes can be used to make composite coatings for car bumpers that better hold their shape in a crash.

o These tubes can also absorb hydrogen, which should enable more efficient storage of future fuels.

o Nanotube enabled space elevators

o You imagine yourself…

Nanotubes are also being used in designing more efficient solar cells in the area of alternative energy (also called clean technology).

Nanotech Energy Applications
Ten Enabling Breakthroughs

1. **Batteries and super-capacitors**. A revolution to improve by 10-100 times for automotive and distributed generation applications.
2. **Direct photoconversion of light** + water to produce H_2
3. **Fuel cells**. A revolution to reduce the cost by nearly 10 to 100 fold.
4. **H_2 storage**. A revolution in light weight materials for pressure tanks and/or a new light weight, easily reversible hydrogen chemisorptions system.
5. **Nanoelectronics** to revolutionize computers, sensors, and devices.
6. **Nanomaterials/ coatings** that will enable vastly lower the cost of deep drilling, to enable HDR (hot dry rock) geothermal heat mining.
7. **Nanotech lighting** to replace incandescent and fluorescent lights
8. **Photocatalytic** reduction of CO_2 to produce a liquid fuel such as methanol.
9. **Photovoltaics**. A revolutionary technology with the possibility of reducing cost by 10 to 100 fold.
10. **Superconductors** with which to rewire the global electrical transmission grid, and also to replace aluminum and copper wires essentially everywhere.

You can go as far as you would like in artificial manufacturing, but it's always useful to look back at nature: the greatest manufacturer and designer →

Learning from Nature

- ◆ Nature performs most of its basic design work at nanoscale.

- ◆ This work mostly stays hidden at nano level and becomes the underlying force in shaping the macro features that we see.

- ◆ Some products of nature also directly exhibit the nano design in the macro world.

- ◆ We can learn from nature's nano design to make nanotech products.

 *Can you please give me an example?*

Learning from Nature:

The Gecko Glue

Figure 3.1 Learning from nature: a gecko

- The gecko can walk up the glass and even hang upside down. **How?**
- The hairs on its feet are so small and elastic that they can exploit forces such as electromagnetic force that pulls molecules together, *sticking* the gecko to the ceiling.
- Learning from this design, nanotechnologists proposed to make sticky tapes or glue that composed of gecko-like synthetic hairs.
- One way of doing it: multiwalled carbon nanotubes added to the surface of a polymer.
- The resulting tape will have adhesive strength about 200 times greater than that of gecko foot hairs: *more power to nanotechnology!*

Summary and Conclusions

♦ Nanotechnology can be used to enhance the quality and characteristics of existing products and also to create new products.

♦ The nanotechnology-enhanced products are already filling the market place.

♦ Nanotechnology products and applications span a broad spectrum from smart clothing to cosmetics to biosensors to carbon nanotubes.

♦ Nanotechnology is helping to design alternative energy solutions, also called *clean technology* or *Cleantech*.

♦ Some nanotechnology products are being created based on our understanding of how nature works at nanoscale.

Self Test Exercises

1 Name some first generation nano products.

2 Nanotechnology has something to do with the evolution of video games from arcade size of the past such as Pong, Frogger, and PacMan to the modern more sophisticated games being played in homes on platforms such as?

3 Name the types of nanotechnology product types in spatial dimensions.

4 These nanoparticles are used in sun creams because they absorb ultra violet (UV) light:

5 Proteins and nucleic acids are used as probe molecules in what kind of sensor applications?

Write Your Own Notes

4

Nanomaterial Characterization

Learning Objectives

1. Nanomaterial Characterization: What and Why?

2. Tools and Techniques of Characterization

3. Spectroscopy

4. Microscopy

5. Transmission Electron Microscope

6. Scanning Electron Microscope

7. Scanning Probe Microscope

8. Scanning Tunneling Microscope

9. Atomic Force Microscope

10. Summary and Conclusions

11. Self Test Exercises

Nanomaterial Characterization: What and Why?

What is material characterization?

 **Material characterization.** the process of studying the structure and properties of a given material by using suitable tools and techniques.

Nanomaterial characterization. Material characterization at nanoscale.

Why study Nanomaterials?

♦ **Scientific Incentives**. Unique, size or scale dependent properties for both exceptional and ordinary applications. Examples:

- Enhanced mechanical properties due to structural perfection and strength

- Making use of quantum properties, for example, quantum dots for LED and lasers

- Surprising but useful catalyst properties exposed through catalyst characterization

- ◆ **Industrial Incentives**. miniaturization and performance:

    - **Moore's Law** has driven semiconductor industry for years to pursue smaller and faster chips, but is running into a dead end.

    - Other industries also moved to nano region as smaller size/scale does offer more possibilities, and it's not a matter of choice.

- ◆ **Developing Capabilities:**

    - Better manufacturing and characterization techniques enable us   to study structures and systems in atomic scale.

    - Use the knowledge from the study to manipulate existing structures and build new structures.

 *How is characterization performed?*

Material Characterization: Tools and Techniques

Recall:

Characterization is the process of studying the structure and properties of a given material by using suitable tools and techniques.

Characterization process may include:

♦ Magnify the specimen to visualize its internal structure.

♦ Collect and analyze the data to determine the distribution of structural components within the specimen, interactions among these components, and other properties.

Tools and techniques for characterization:

♦ **Microscopy.** For visualization
♦ **Spectroscopy.** For analysis

Microscopy. Mainly three kinds:

♦ **Optical microscopes**. No good for visualizing nanostructures.
♦ **Electron microscopes.** Make use of electron waves unlike EM waves used by optical microscopes.
♦ **Atomic force microscopes**. Uses electric force.

 *What is spectroscopy?*

Spectroscopy

Spectroscopy. Technique that uses the interaction of light with matter to study the structure of matter.

Spectroscopy

♦ Uses **the interaction of light with matter** to study the structure of matter

♦ **Modern spectroscopy** also includes non-electromagnetic (non-light) waves such as sounds waves and electron waves.

♦ **Spectrometer**. A tool, system, and a setup used in spectroscopy

♦ **Sample.** The material under study.

♦ Depending on what kind of light is used and how it interacts with the sample, there are several *types* of spectroscopy:

  • Infrared (IR) Spectroscopy
  • Raman Spectroscopy
  • X-Ray spectroscopy
  • X-ray diffraction spectroscopy

Can you please present an example to illustrate spectroscopy?

Spectroscopy: An Example

It's a rather long but thorough example; so grab your favorite drink:

Fact: An aspirin tablet contains a certain amount of acetylsalicylic acid (ASA), one of the most commonly used drugs on this planet.

Goal: To find the amount of ASA (concentration) in a commercial aspirin tablet.

Principle: Absorbance of light, A, by a solution (measured by a spectrometer called spectrophotometer) is proportional to the concentration (strength), C, of the solution :

$$A = kC$$

Where k is the proportionality constant.

Preparation:

1. Prepare a few samples of ASA solutions with known (different) concentrations. These samples will be used to calibrate the spectrometer.
2. Dissolve one tablet of commercial aspirin in a suitable solvent and make 5 liter of solution. We will use the spectrometer and the calibration plot to measure the concentration of ASA in this solution in order to determine the mass of ASA in the tablet.

Steps:

1. Make samples of solutions with known concentrations of ASA in it.

2. Use the spectrometer to obtain absorption spectrum by using the sample with maximum concentration. This means taking the absorbance measurements corresponding to different wavelengths.

3. From the absorption spectra, determine the wavelength, λ_{max} at which the solution absorbs maximum light.

4. Set the spectrometer to λ_{max} . This means it will only let the light of this wavelength to pass trough the solution.

5. Measure the absorbance of all the samples at known concentrations at this wavelength, λ_{max}, and plot absorbance versus concentration.

6. Now measure the absorbance of the solution with unknown concentration, and determine the concentration by comparing the measured absorbance to the graph obtained in step 5.

Absorption spectra (step 2) measured with the spectrometer for an ASA solution sample with known concentration:

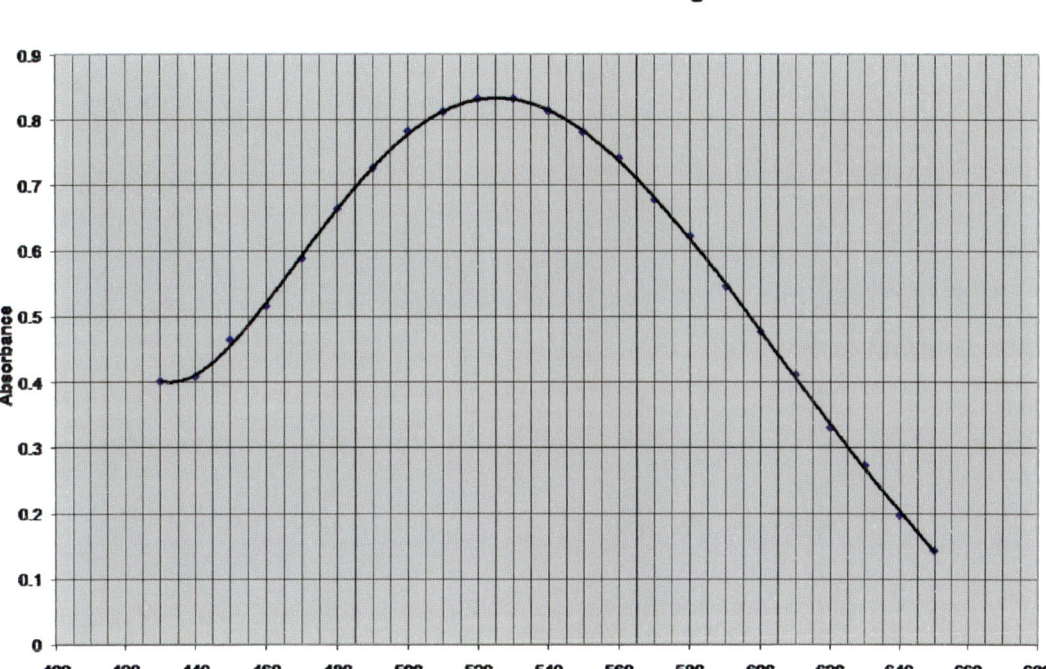

Figure 4.1 Absorption spectra for an ASA solution with a known concentration.

Absorbance is maximum at:

$$\lambda_{max} = 525 \ nm$$

Set the spectrometer to this wavelength for the rest of the experiment.

Use spectrometer to measure the absorbance of ASA solution samples with known concentrations (step 5):

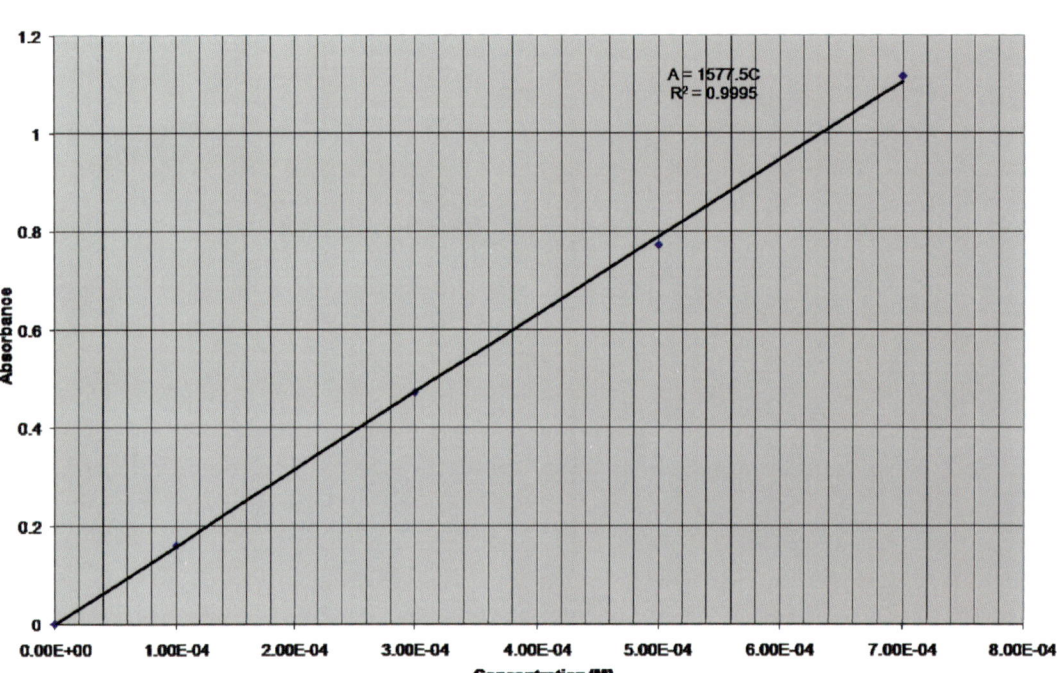

Figure 4.2 Calibration plot for ASA: Absorbance measured against known concentrations

From the calibration plot (Figure 4.2):

```
A = 1577.5 C
```

→

```
C = A/1577.5
```

Analysis:

From the calibration plot:

A = 1577.5 C

⇨ C = A/1577.5

Absorbance, A, for the commercial aspirin solution with unknown concentration was measured to be 0.559.

=>

C = 0.559/1577.5

= 3.54×10^{-4} M

where M is concentration in molarity, that is, the number of moles per liter. Therefore number of moles of ASA in 5 liters of solution can be calculated as:

Moles of ASA = (3.54×10^{-4} moles/L) x 5 L = 1.77×10^{-3} moles

This whole solution was made from one tablet, and therefore these moles were in the tablet. The mass of one mole of ASA is 180.15 g. Therefore the mass of ASA in the tablet can be calculated as:

Mass of ASA in the commercial aspirin tablet = moles of ASA in the tablet x molar mass of ASA

```
= 1.77x10⁻³ moles x 180.15 g/mole = 0.319g = 319 mg
```

 *What is IR spectroscopy?*

IR Spectroscopy

IR spectroscopy. The spectroscopy that uses the infrared radiation (IR) part of electromagnetic (EM) spectrum to determine the composition of a specimen. Can be used to study a specimen at molecular level.

Goal. To find the frequency of vibration of the bonds of atoms in molecules.

Trick. The molecules of the specimen will absorb the light whose frequency (or wavelength) matches with that of the atomic bonds in the molecules.

Process.

1. Spectrometer puts the IR light on the specimen.
2. The IR light shines on a vibrating molecule.
3. From this light, the molecule absorbs only that part whose frequency matches with the frequency of the molecular vibration.
4. The spectrometer detects the frequency of the light component absorbed by the molecules by measuring the light that was not absorbed and therefore came out. This reveals the **composition of the specimen**.

So, to study the composition of a compound, you use IR spectroscopy; but to actually study the chemical bonds in a compound, you need to use Raman spectroscopy.

What is Raman spectroscopy?

Raman Spectroscopy

 **Raman spectroscopy**. The spectroscopy that is based on the study of the phonons (vibrations of the material structure) and photons emitted by the atoms after they absorb the light incident on the specimen.

Here is how it typically works:

1. A specimen is illuminated with a laser beam.

2. The incident photon is absorbed by an atom that goes to an excited state as a result.

3. The excited atom comes to a lower energy state by emitting a photon, phonon, or both. The emitted photons are called scattered photons.

4. The frequency (or wavelength) of the scattered photons provides information about the chemical bonds of the molecules in the specimen material.

Example of Use:

- Study the various crystallographic forms of carbon

Much like photons are quanta of light, phonons are quanta of vibrations, for example, the vibrations occurring in the atomic lattice of a material.

 *So, by using Raman spectroscopy, you can study the chemical (atomic) bonds in compounds and molecules. But if you want to probe further to study the electronic structure of atoms in the sample, you need to use X-ray spectroscopy.*

 *Now, What is X-ray spectroscopy?*

X-ray Spectroscopy

X-ray spectroscopy. The spectroscopy that uses the X-ray part of electromagnetic (EM) spectrum to determine the electronic structure of a material. X-rays are the high energy, that is, high frequency and low wavelength (in the range of 10 nm to 0.01 nm) EM radiation. **There are different types of X-ray spectroscopies:**

- X-ray photoelectron emission
- X-ray diffraction

X-ray photoelectron emission:

1. X-rays are focused on the specimen.

2. The energetic X-ray photons knock out electrons from the atoms in the specimen material.

3. A knocked out electron, called photoelectron, has the kinetic energy given by the following equation:

$$E_k = hf_p - E_i$$

where f_p is the frequency of the incident photon (in the X-ray), and E_i is the ionization energy required to knock out the electron from the material.

4. A detector detects the emitted photoelectrons, and the energy distributions of these electrons are measured.

5. The energy distributions of the emitted electrons reveal information about the molecules and atoms in the specimen material.

What is X-ray diffraction?

X-ray Diffraction Spectroscopy

X-ray diffraction spectroscopy is the spectroscopy that uses diffraction of X-rays through the target material to determine the internal structure of the material.

Process:

1. A collimated (lined up) beam of X-rays is directed at the crystal.

2. The X-ray photons in the incident beam are scattered in all directions from different layers of atoms in the crystals.

3. Due to the regular arrangement of atoms in the crystal, in certain directions, the scattered waves interfere with one another constructively. These waves can be measured.

The scattered waves with constructive interference (that are measured) are those waves which are parallel to each other.

OK, let's backup a little; what is diffraction, anyway?

Diffraction

 **Diffraction**. Bending of waves around obstacles in their path:

♦ Diffraction effects are generally most pronounced for waves whose wavelength matches with that of the diffracting objects.

♦ Different parts of a wave travelling to the observer by different paths interfere with one another and give rise to complex intensity patterns.

 *There are so many types of spectroscopy. Can you please summarize them in a table?*

Summary of Spectroscopy

Table 4.1 Characteristics of different types of spectroscopy.

Spectroscopy Type	Basic Operational Principle	Studied	Used To Determine
IR	*Photons in.* Light from the IR part of the EM spectrum is focused on the specimen and the absorption spectra is studied.	Absorption spectra	Composition of a compound
Raman	*Photons in, photons and phonons out.* The light is absorbed by the specimen and the photons emitted by the specimen are studied	Vibrational modes in the specimen	Chemical bonds among atoms and among molecules
X-ray photoelectron emission	*Photons in, electrons out.* High energy photons (X-rays) knock out electrons from the atoms of the specimen, whose kinetic energy is measured.	Kinetic energy of electrons knocked out by the X-rays from the specimen	Electronic structure of atoms in the specimen
X-ray diffraction	*Photons in, photons out.* X rays are scattered/diffracted from the atoms of the target material	The spectra of the scattered/diffracted X-rays	Internal structure of the material such as the position of atoms in a crystal.

Microscopy

Microscopy is the science of observing and investigating small objects and structures by using instruments such as microscopes.

Different kinds of microscopes:

♦ Optical microscope

♦ Scanning electron microscope

♦ Transmission electron microscope

♦ Atomic force microscope

Difference between magnification and resolution:

♦ **Magnification.** The factor by which the image of an object is enlarged as compared to the object size.

♦ **Resolution.** The smallest size of the object that can be distinguished from its surrounding material. Determines how much detail of the structure can be observed.

Optical microscope, resolving power limited by the frequency of wavelength of visible light:

Average wavelength of white light $= 500$nm.

$$P_R = \frac{\lambda}{2 N_A} = \frac{500 nm}{2 * 1.25}$$

$$= 200 nm$$

$$= 0.200 \mu m$$

If you need better resolution, you need to use electrons rather than photons for visualization! ➜

Seeing Nanostructures with Electron Waves

♦ According to the wave-particle duality of matter (in quantum mechanics), electrons can exhibit wave properties.

♦ Electron waves can have smaller wavelengths than the smallest wavelength of photons (light) ➔ Better resolution with electron waves.

♦ Depending on how the electrons interact with the sample, several microscopes are built based on this and related quantum mechanical effects:

- Transmission electron microscope

- Scanning electron microscope

- Scanning probe microscope

- Scanning tunneling microscope

- Atomic force microscope

 *What is transmission electron microscope?*

Transmission Electron Microscope

 **Transmission electron microscope**. An electron microscope which uses the transmission of electrons through the target material to take the image of the target specimen.

The basic idea is that the electrons transmitting through the specimen will be scattered by the constituent particles of the specimen, and the scattering will expose the internal pattern (or nanostructure) of the specimen, which will be captured in an image.

Process:

1. A source such as an electron gun generates a beam of electrons.

2. By using electromagnetic field, the electron beam is accelerated and focused to the target specimen.

3. The incident electrons are scattered by the internal structure of the specimen as they transmit through the specimen.

4. The scattered electrons, after they transmit through the specimen, are focused by objective lens and amplified by magnifying lens to generate an enlarged image of the specimen's structure.

5. The image strikes a phosphor screen, which generates light, which in turn enables a user to see the image.

Example: A darker area of the image represents a thicker (denser) part of the specimen that did not allow many electrons to pass through it. In contrast; a lighter area of the image represents a thinner (or less dense) part of the specimen that allowed many electrons to pass through it.

 *Can you please show an image taken by TEM?*

TEM Image: An Example

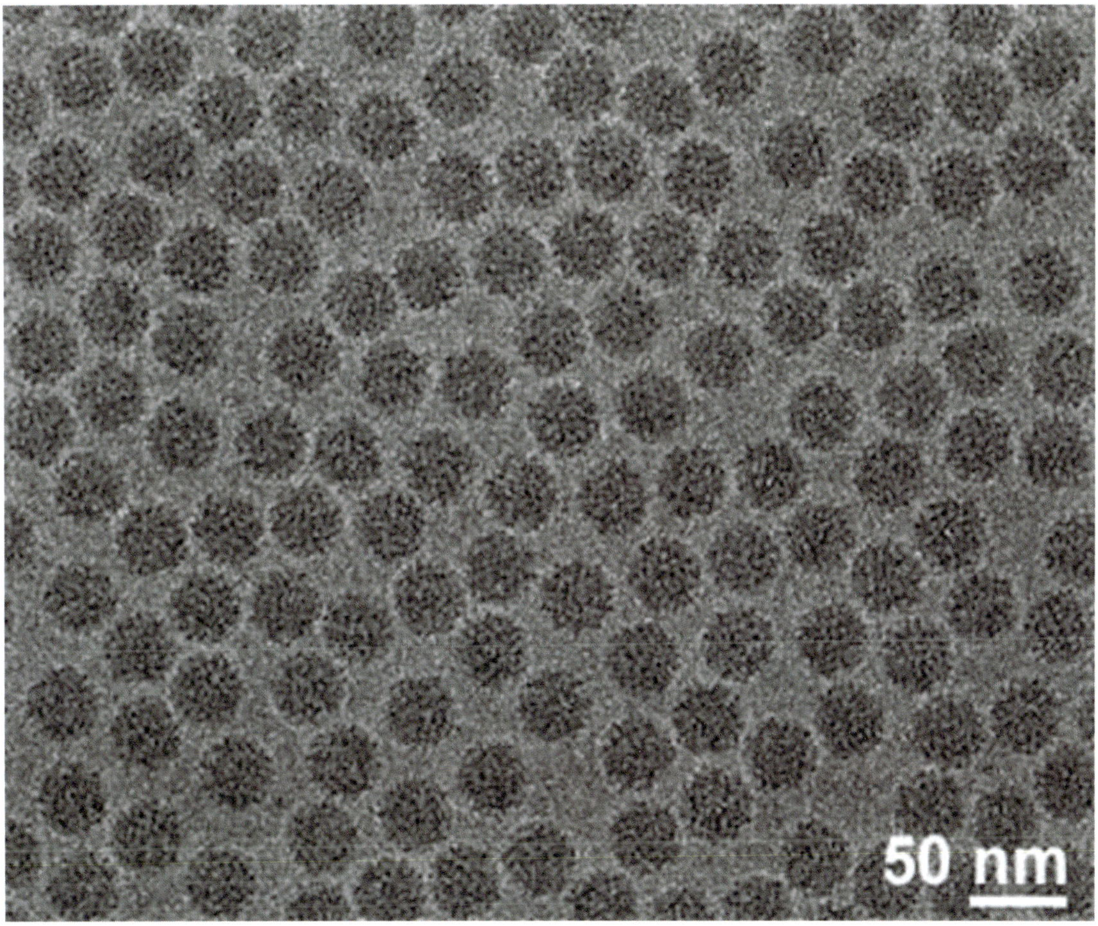

Figure 4.3 TEM image of tomato bushy stunt virus, called TBSV virus.

What is scanning electron microscope?

Scanning Electron Microscope

Scanning electron microscope is the microscope that observes the surface of a bulk object by scanning the surface with an electron beam and measuring the properties of what comes out of the specimen as a result of scanning.

Process:

1. A source such as an electron gun generates a beam of electrons.

2. By using electromagnetic field, the electron beam is accelerated and focused toward the target specimen.

3. The electron beam passes through pairs of scanning coils in the objective lens, which deflect the beam in a raster fashion over a specimen surface for scanning.

4. The electrons in the beam, called primary electrons, knock out the electrons from the atoms in the specimen. These knocked out electrons are called secondary electrons or secondary particles. Some primary electrons will also come back out of the specimen and are called backscattered electrons.

5. The secondary electrons that escape the specimen surface are observed by a detector such as a scintillator-photomultiplier.

6. The detector counts and measures the secondary electrons and sends the information such as intensity to the amplifier, and this information is rendered into an image.

7. Some X-ray photons can also be produced from the interaction of the incident beam with the sample.

Can you show an image taken by SEM?

SEM Images: An Example

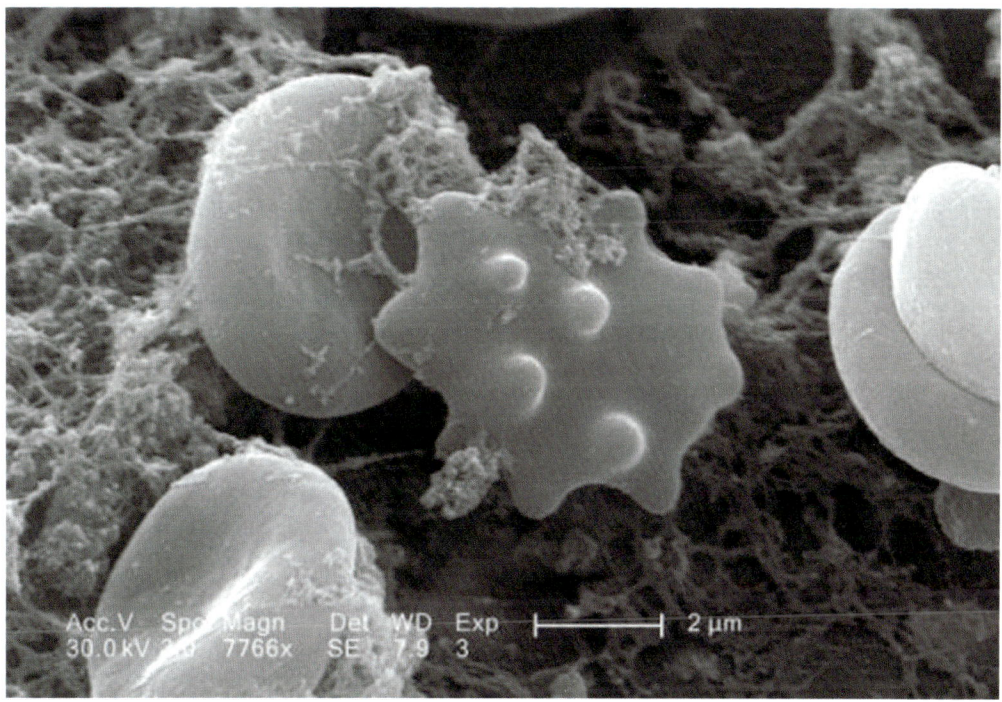

Figure 4.4 A SEM image that depicts a number of red blood cells found enmeshed in a fibrinous matrix; Magnification: 11397x.

 *What is scanning probe microscope?*

Scanning Probe Microscope

 **Scanning probe microscope** is a microscope that generates an image of a specimen's surface by using a physical probe to scan the specimen.

An image of the surface is obtained by mechanically moving the probe in a raster-like scan of the specimen, line by line, and recording the interaction of the probe-surface as a function of position.

Process:

1. A probe tip is fixed or mounted at the end of a tiny beam.

2. The tip scans and records the surface of the specimen by touching it, as the surface is moved underneath it in a grid pattern.

3. The tip detects the changes in the surface such as height and electrical properties. This is facilitated by measuring up and down movements by a laser beam, and by measuring the current between the surface and the tip. The current is produced by the voltage difference between the surface and the tip.

4. The measurements in step 3 are sent to an optical detector that creates an image of the surface.

 *What is scanning tunneling microscope?*

Scanning Tunneling Microscope

 **Scanning tunneling microscope** is a microscope that uses a probing tip of the atomic size to scan the surface of a specimen.

Here is how it works:

1. A voltage is applied between the probe tip and the surface.

2. The surface is scanned by keeping the tip at an atomic distance (e.g. 0.2 nm) away from the surface.

3. Due to the quantum mechanical effect called tunneling, some electrons flow from the surface to the probe, hence generating a current called tunneling current, even if the probe is not directly touching the surface.

An STM can achieve a resolution to observe single atoms, that is, of the order of 0.2 nm.

An STM can be used for the following purposes:

♦ Characterizing the roughness of a surface

♦ Observing the surface defects

♦ Determining the size and conformation of the molecules on the surface

♦ Helping to see and position individual atoms

♦ Studying DNA molecules

 *Can you please give an example of STM images?*

STM Images: An Example

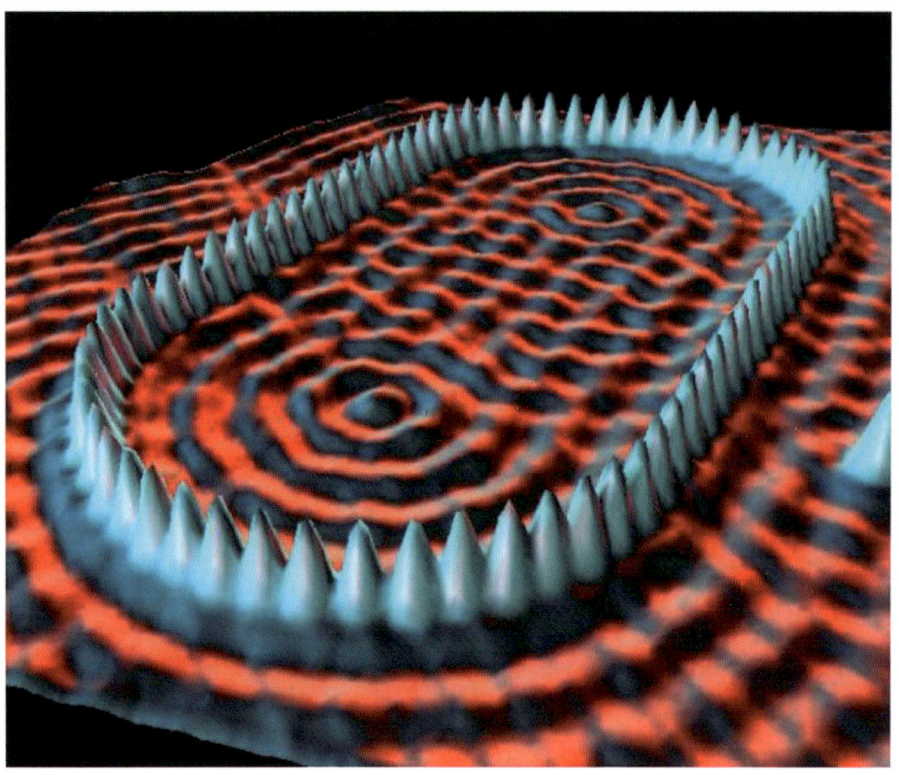

Figure 4.5 Iron atoms individually positioned and imaged by the STM

 *What is atomic force microscope?*

Atomic Force Microscope

 **Atomic force microscope (AFM)** is a microscope that uses a probing tip of the atomic size to scan the surface of a specimen.

♦ AFM It works just like STM except that it uses the atomic force between the tip and the surface instead of tunneling current to scan the surface.

♦ The scanning device is a microscale cantilever with a sharp tip at its end.

♦ The scanning works in the following way:

1. When the tip of the cantilever is brought in the proximity of the target surface, the atomic force between the tip and the surface deflects the cantilever.

2. The force depends on the distance between the tip and the surface.

3. As the surface is scanned with the tip, the tip is moved up and down to keep the force constant. In other words, the ridges and valleys on the surface will cause the up and down movements of the tip, which in turn will deflect the cantilever.

4. The deflections are measured by a laser beam reflected from the back of the cantilever on to a position-sensitive photodiode. This arrangement is also called an optical lever.

 *Can you please give an example of manipulating atoms?*

Manipulation of Atoms: An Example

Demonstrating the manipulation power of STM and AFM at atomic level:

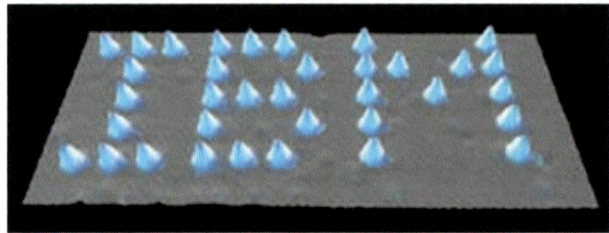

Figure 4.6 The letters "IBM" spelled in xenon atoms, and imaged by the atomic force microscope.

The story goes like this:

1. In 1989, an IBM researcher used STM to move individual atoms of Xenon gas.

2. The gas was cooled to extremely low temperature so that the gas atoms sat still.

3. By manipulating the atoms, the researcher was able to move atoms to spell IBM on a nickel surface.

4. The AFM was then used to capture the image.

 *You have mentioned so many microscopes. Can you please summarize them at one place?*

Characteristics of some commonly used microscopes

	Transmission Electron Microscope (TEM)	Scanning Electron Microscope (SEM)	Scanning Tunneling Microscope (STM)	Atomic Force Microscope (AFM)
Typical resolution	0.1 nm – 0.2 nm	1 nm – 20 nm	0.2 nm	≤ 0.1 nm
Basic operational principle	Electrons in, electrons out. An electron beam is transmitted through the specimen and the scattered electrons are measured	Primary electrons in, secondary electrons out. A beam of electrons knocks out electrons in the specimen which escape the surface and are imaged	The current between the scanning probe and the specimen surface is monitored	The atomic force between the scanning tip and the specimen surface is monitored
Main advantage	Excellent resolution	Ability to image relatively large area of the specimen	Three dimensional images of metallic surfaces at atomic scale	Three dimensional images at atomic scale
Main disadvantage	Field of view is relatively small	Resolution not high enough to image down to atomic scale	Only useful for conductive surfaces	Image area small: few micrometers squared, for example, (150 μm x 150 μm)
Year of original invention	1938	Developed: 1942. Commercially released: around 1965	1981	1986

Summary and Conclusions

♦ It's important to study the existing structures in order to build new structures. Nanotechnology tools help to study nanostructures.

♦ Spectroscopy and microscopy are the major techniques used to study the structures of materials, that is, material characterization.

♦ Microscopy is used for visualization and spectroscopy for the analysis.

♦ Some common microscopes used for the characterization of nanomaterial are:

  o Transmission Electron Microscope (TEM)

  o Scanning Electron Microscope (SEM)

  o Scanning Probe Microscope (SPM)

  o Scanning Tunneling Microscope (STM)

  o Atomic Force Microscope (AFM)

♦ Different microscopes vary by the operational principles and offer different resolutions.

Self Test Exercises

1 Why can't you use the optical microscope to study nanostructures?

(Hint: resolution and wavelength)

2 Rearrange the entries in the second column of the following table to correct errors:

Technique	Feature
Microscopy	Used for analysis
Spectroscopy	Used for visualization
Optical microscope	Uses Electromagnetic force
Electron microscope	Uses electromagnetic waves
Atomic force microscope	Uses electron waves

3 Rearrange the entries in the second column of the following table to match them correctly with the entries in the first column.

Type of Spectroscopy	Basic Operational principle
IR	Photons in, electrons out
X-ray diffraction	Photons in, phonons out
X-ray photoelectron emission	Photons in
Raman	Photons in, photons out

4 Rearrange the entries in the second column of the following table to match them correctly with the entries in the first column.

Type of Microscope	Basic Operational principle
TEM	Primary electrons in, secondary electrons out
SEM	Electrons in, electrons out.
STM	The current between the scanning probe and the specimen surface is monitored
AFM	The atomic force between the scanning tip and the specimen surface is monitored

5 Which microscope uses the principle of quantum mechanical tunneling?

5

Nanofabrication

Learning Objectives

1. Definitions

2. Matter and Energy

3. Human History and Material

4. Fabrication Terminology

5. Nanofabrication Techniques

6. Nanolithography

Defining Fabrication

 **Definitions**

Fabrication. A process used to create components, devices, and products, for example chips and integrated circuits.

Fabs. Specilized facilities in which the entire manufacturing process from start to packaging is performed.

Nanofabrication. Fabrication of nanostructures, that is, the structures with at least one lateral dimension within the nanoscale. It includes a set of different techniques such as nanolithography and nanoimprint.

In order to understand fabrication, it's important to understand some aspects of materials such as mass and energy ➔

Mass and Energy

➢ Look at the world (or the universe) around you.

➢ It's made of matter and energy. For example, the car you drive is matter, and it needs energy produced by the oil (matter) to drive.

➢ Matter and energy are equivalent to each other and therefore inter-convertible, and this relationship is expressed in Einstein's famous equation:

$$E = mc^2$$

where

- c is the speed of light

- m is the effective mass of a particle given by the following equation:

$$m = \frac{m_0}{1 - \dfrac{v^2}{c^2}}$$

where

- m_0 is the mass of the particle when the particle is at rest and is called rest mass

- v is the speed of the particle in motion.

You can see from this equation that the mass of a particle increases when the particle is moving. This is part of the **special theory of relativity**.

Energy Units

Energy Units:

$$\text{Joule} = \text{Nm} = \text{Kg} \, (\text{m/s})^2$$

$$1 \ \text{eV} = 1.60 \text{x} 10^{-19} \ \text{J}$$

$$1 \ \text{MeV} = 1 \text{x} 10^{6} \ \text{ev} = 1.60 \text{x} 10^{-13} \ \text{J}$$

where:

eV = Electron volt

MeV = Mega electron volt

Matter and Material
Big Picture

- Nucleons (protons and neutrons) are made of quarks.

- Nuclei are made of nucleons.

- Atoms are made of nuclei and electrons (an example of a class of fundamental particles called leptons).

- Molecules are made of atoms.

- All materials consist of molecules.

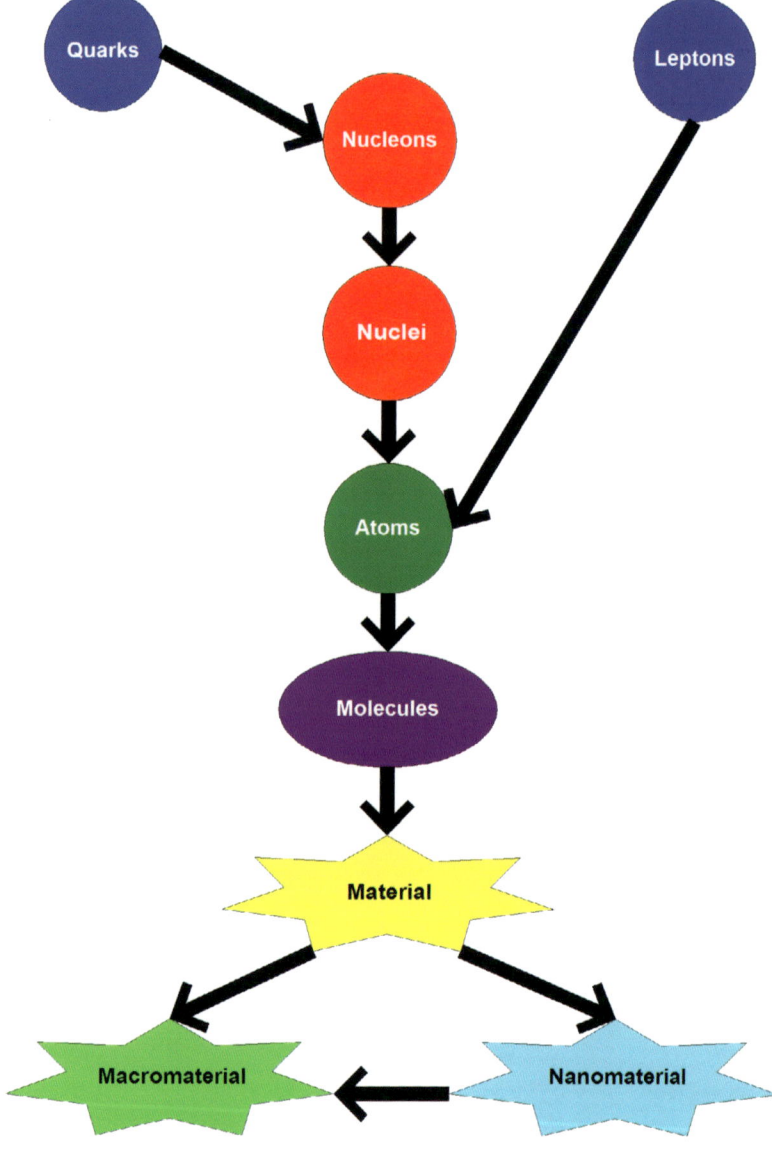

Figure 5.1 Building blocks of material

Matter is anything that has mass and occupies space. That means all particles that has mass are matter as well.

Human History: The History of Materials

- ➢ **Stone age.** First stage of a prehistoric time period when human used tools made of different kinds of stone.
- ➢ **Bronze age**. A period in human history marked by the making and using of bronze: a broad range of copper alloys with tin and other elements. This was the second stage of prehistoric period.
- ➢ **Iron age.** A period in human history when iron was the main ingredient in the tools used by human. It's the third and final stage of prehistoric period
- ➢ **Industrial age.** Driven by engines and machines, the single most important factor of this age is the large scale production of things, that is, manufacturing.
- ➢ **Information age.** Current age, fundamentally driven by semiconductors and microchips.
- ➢ **Molecular age/Nano age**. The age we are entering, driven by our understanding and manipulation of material structures at nanoscale such as moleculcs and atoms.

Nanomaterial: Types

 **Nanomaterial.** Any material that has morphological features of nano size at least in one dimension.

You can categorize nanomaterial based on size or its response to change.

Size:

♦ **Bulk Nanomaterial.** Bulk material with nanoparticles in it that give it some unique characteristics. Examples: Stain removers in clothing, water resistant wind shields, sheets with built-in self-clean capabilities, and so on.

♦ **Nano Nanomaterial.** Nanomaterial at nanoscale. Examples, buckyballs and carbon nanotubes.

Change:

♦ **Static smart materials**. These materials always behave in a certain way and resist to any change. For example a shirt with a built-in no-stain capability will stay stainless.

♦ **Dynamic smart materials**. These are the materials that react to the external forces in such a way so that they can change their properties. For example, a polymer material coated on a vehicle changes color in response to external forces such as change in weather or driver's instructions.

Nanomaterial: Smart Material

Smart in Which Ways?

Some examples:

> **Recognition and response**. The capability to recognize the external conditions such as chemical or electromagnetic stimuli and respond in a certain way. Nanotubes, for example, contain the capability of recognition.

> **Separation**. The capability of performing separation operation on the substance to which the material is exposed. For example, as you digest food, your digestive system separates the food elements that have nutritional values from those that do not. The separation in nanostructures, for example, can be implemented by a technique that permits small molecules to pass through nanosize pores whereas blocking the larger molecules.

> **Self healing**. The simplest self-healing structure is the structure that responds to local breaks in a continuous fabric by repairing these breaks. Think of a punctured tire repairing itself immediately after the puncturing event occurs.

Fabrication: Terminology

➢ **Substrate**. The material on which a process is conducted. In electronics, it is the material on which semiconductor devices are fabricated. In this context, it's also called a wafer.

➢ **Wafer**. In general, it is a small, thin, and crisp cake or biscuit. In electronics, a wafer is a small and thin disk of semiconductor material, such as silicon, on which an integrated circuit is formed. It's also called a silicon wafer.

➢ **Resist**. A thin layer, usually made of polymers, used to transfer a circuit pattern to the semiconductor substrate such as silicon wafer on which the resist is deposited.

➢ **Spin-coating**. A procedure used to spread a uniform thin film on a flat substrate such as a silicon wafer. An excess amount of solvent is placed on the substrate, which then is spinned at high speed by using a machine called spin coater or spinner. The solvent is spread by the centrifugal force caused by spinning.

➢ **Mask**. Also called photomask in photolithography; it is an opaque plate with holes in it to allow light shine through in a pre-determined pattern.

➢ **Etching**. The process of creating a design in the metal by cutting into the unprotected parts of the metal by using acid.

   *What is nanofabrication?*

Nanofabrication

Two approaches

Two approaches to fabricating nanostructures and nanodevices:

Top-Down. Large object are modified to build nanostructures.

Examples:

- Nanolithography
- Nanoimprint
- Thin-film deposition and growth
- Etching technology
- Scanning probe
- Mechanical polishing

Bottom-Up. Smaller building blocks are assembled into nanostructures.

Examples:

- Chemical synthesis
- Laser trapping
- Self assembly
- Colloidal aggregation.

 *What is Nanolithography?*

Nanolithography

Lithography

- ♦ Invented by Alois Senefelder in 1798.
- ♦ A technique for printing on a smooth surface. It can be used to print text or artwork onto paper or another suitable material.
- ♦ In electronics, it's used for manufacturing semiconductor and MEMS (Microelectromechanical Systems) devices.

Nanolithography. A set of lithographic techniques used to structure material on a nano scale.

Examples of Nanolithography

- ♦ Electron beam lithography
- ♦ Nanoimprint.

Can you please tell me more about electron beam lithography and nanoimprint?

E-beam lithography and Nanoimprint

E-beam Lithography

- Uses an electron beam with the width of the order of nanometers to expose an electron-sensitive resist.
- Electron wavelength of 0.1 nm can be achieved ➔ Excellent resolution, about 10 nm.
- Resolution is limited by the electron scattering in the resist.

Drawback:

Low throughput because it works in a serial fashion.

Nanoimprint:

1. Use e-beam lithography to create a stamp of hard material such as silicon.
2. A resist-coated substrate is stamped.
3. Resist residues in the stamp are removed.
4. Either the substrate is etched, or in case of a metallic nanostructure, the metal is evaporated and a lift-off is performed.

 *E-beam lithography and nanoimprint are examples of top-down fabrication methods; another example is scanning probe technique.*

 *What is scanning probe technique?*

Scanning Probe Techniques

 Scanning probe techniques are the techniques that use the scanning probe tip of scanning probe microscopes such as scanning tunneling microscope (STM) and atomic force microscope (AFM).

Examples

Scanning Probe Oxidation. Create oxidation by operating scanning probe of the microscope in air and bias it at sufficiently high voltage, say -2 to -10 V. Following are the main steps of the process:

1. The surface of the silicon wafer is passivated with hydrogen atoms by dipping it, e.g., in hydrogen fluoride solution.

2. The passivated surface is oxidized.

3. Patterns of oxide are written on the silicon surface, which work as a mask for dry or wet etching.

Achieved line width of the pattern around 10 nm

Scanning Probe Resist Exposure. Use electrons emitted from a scanning probe tip of a microscope to expose an electron-sensitive resist. Following are the main steps of a typical process:

1. The silicon wafer is cleaned to remove the native oxide with a dip in the HF solution.

2. A resist is then spin-coated onto the surface of the wafer up to a thickness of 35 to 100 nm.

3. The scanning probe tip of the microscope is moved over the surface to achieve exposure.

4. After the exposure, the resist is developed in an appropriate standard solution.

Achieved line width of features about 50 nm

Dip-Pen Nanolithography. Use the tip of an AFM to write the pattern:

1. The tip of an AFM is inked with a chemical, for example, by dipping the tip in a solution that contains low concentration of the desired molecules. The dipping can be followed by a drying step.

2. The tip is brought into contact with the surface.

3. The molecules flow from the tip to the surface.

Achieved line width of the pattern about 10 nm

Thin film deposition methods are also examples of top-down approach.

What are thin film deposition methods?

Thin Film Deposition Methods

 **Thin film deposition.** Any technique used to deposit a thin film of material on a substrate or on a previously deposited layer.

Two Broader Types of Thin Film Deposition:

Physical deposition. The process involved in the deposition is a physical process, that is, no chemical reaction is involved.

Chemical deposition. The deposition method involves the process that includes chemical reaction.

Physical Deposition Examples:

- Cathodic arc deposition
- Pulsed laser deposition
- Sputtering
- Thermal evaporator

Chemical Deposition Examples:

- Chemical vapor deposition (CVD): many variations

 *How does physical deposition work?*

Physical Deposition

Typical physical deposition process involves three components:

1. **Source**. The material that would be deposited is staged in an energetic environment so that the material particles escape the surface.

2. **Sink**. Facing the source is a cooler surface that sucks the energy out of the arriving particles, and as a result the particles cool down and deposit and form a solid layer on the surface which is usually a substrate.

3. **Vacuum chamber**. The whole system is contained in a vacuum chamber that allows the particles to travel freely without interacting with other particles or matter.

 *Can you please describe some examples of physical deposition methods?*

Physical Deposition Methods

♦ **Cathodic arc deposition**.

  o Vapor deposition technique.

  o Particles are the ions blasted from a cathode with an extremely powerful electric arc.

  o A reactive gas interacts with the ion flux to deposit a compound film.

♦ **Pulsed laser deposition.**

  o A high power laser beam is directed at the source material that converts the converts the material to plasma.

  o The particles in the plasma evaporate and deposit as a thin film on a substrate such as a silicon wafer facing the source.

♦ **Sputtering**.

  o Atoms are ejected (sputtered) from a solid material by bombarding it with energetic ions.

  o The atoms are ejected due to the momentum transfer between the ions and the atoms in the material, and evaporation is not involved.

  o So, the source material from which the atoms are ejected can be kept at a low temperature.

- o For the sputtering to happen the incident ion must reach or exceed a minimum energy value called threshold energy which is equal to the binding energy of the atoms that binds the atoms to the surface.

- o The threshold energy is somewhere in the range of 10 to 100 eV.

- o The atoms eroded from the material such as SiO_2 then deposit on a substrate such as silicon wafer to make the thin film.

♦ **Thermal evaporator**:

- o Use the heat from electric resistance heater to melt the material and raise its vapor pressure to reach or exceed a threshold value.
- o The evaporated particles produced this way then travel in the vacuum chamber toward the substrate where they deposit to make the film.

 *Can you please elaborate on chemical deposition?*

Chemical Deposition

Chemical change is a key component of chemical deposition methods:

Chemical vapor deposition (CVD)

1. A reactive (volatile) substance called precursor is used as the source material.
2. The source material reacts (or decomposes) on the surface of the substrate to produce the desired deposit.
3. The undesired byproduct of the chemical reaction is removed by using an appropriate method such as the gas flow through the chamber.

Several different CVD methods

Differentiated by:

✓ Type of the chemical reaction
✓ Process conditions

Because nanotechnology probes deeper to molecular and atomic level, the role of atoms and corresponding elements becomes more apparent and significant. The element that plays the most significant role is: Carbon?

What's special about carbon?

What's Special about Carbon?

Carbon: The heart of organic matter**.**

- ✓ More than 25 million classified chemical compounds, and about 95% of them are organic.
- ✓ The uniqueness of carbon atoms lays in the ability to bond strongly to one another to form chains of different sizes.

Reasons for Enormous number of carbon compounds:

- ♦ The small size of the carbon atom contributes to its ability to make bonds with other small atoms including itself.

- ♦ The carbon atom has four valence electrons. It gives it the ability and flexibility to make bonds with at maximum four other atoms in various ways. For example, carbon atoms can bond to make:

    - o Short chains, which will give you a gas.

    - o Long chains, which will give you a solid such as plastic

    - o A 3-dimensional lattice, which will give you a hard material such as a diamond

- ♦ The availability of four electrons to make bonds enables the carbon atom to leverage the *octet rule* for making stable molecules.

Can you please give me an example of a material purely made of carbon?

Graphite

- ♦ **Examples:** pencils, electrodes of an electrical arc lamp.
- ♦ Purely made of carbon molecules
- ♦ Each molecule consists of six carbon atoms in the form of a ring.
- ♦ Each carbon atom covalently bonded to three other carbon atoms using three of its four valence electrons.
- ♦ Each carbon atom still has one valence electron unused. This makes graphite conductive.

Two basic properties of graphite:

- ✓ Conductivity
- ✓ Strength

Graphite sheets folded in certain ways make:

- ✓ Buckyballs
- ✓ Nanotubes

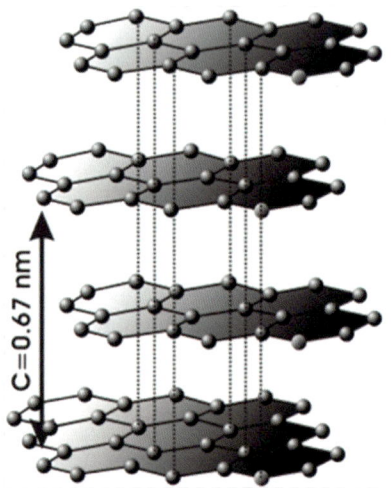

Figure 5.2 Illustration of the graphite structure

 What are buckyballs?

Buckyballs

- **Buckyball,** a carbon molecule composed of 60 carbon atoms in the form of a hollow sphere: C_{60}.
- **Diameter**. 1 nm
- Each molecule consists of 20 six-member rings and 12 five-member rings.
- Each carbon atom covalently bonded to three neighboring carbon atoms.

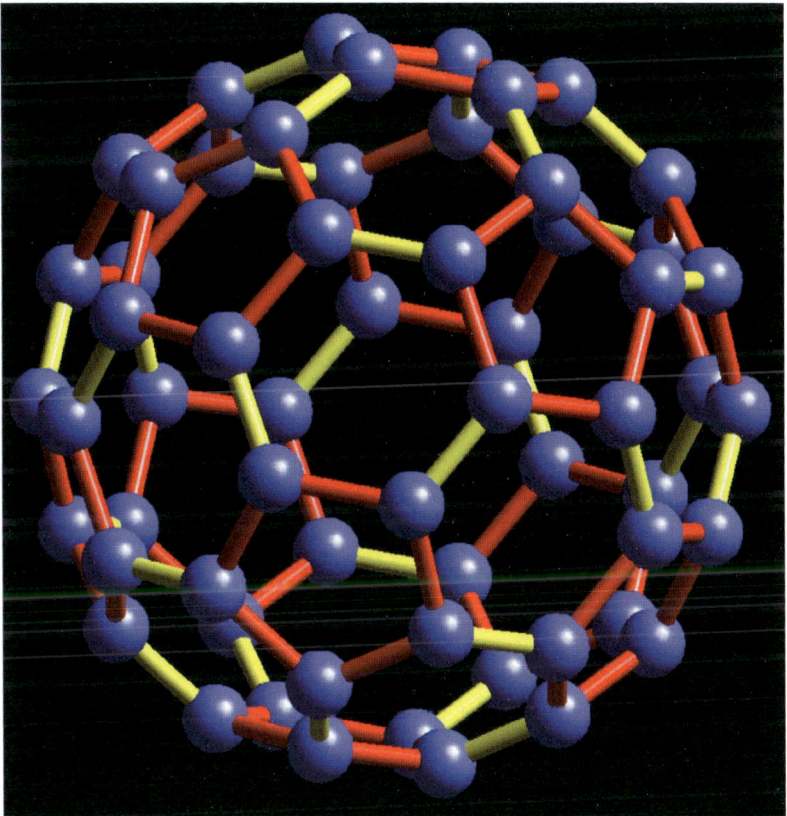

Figure 5.3 Illustration of a buckyball

 *Multiple buckyballs are held together by inter-molecular force called van der Waals force, a form of electromagnetic force.*

Buckyballs: Applications

Useful Characteristics:

- ◆ Each atom has one unused valence electron => Conductivity
- ◆ Strength
- ◆ Large structure, yet small size.
- ◆ Empty space inside the ball ➜ Can be used to transport things

Uses:

- ◆ Use in armors (*strength*)
- ◆ Facilitate superconductivity (*free valence electron*)
- ◆ Dope metal inside the buckyball and use it for medical imaging
- ◆ Medical uses; carry other molecules (*hollowness*):
    - o Bind certain antibiotics to buckyballs and target undesired cells such as cancer cells.
    - o Deliver drugs directly to the infected part of the body
    - o Use buckyballs to neutralize free radicals in the body

Free radical is a molecule that has an unpaired electron, which makes this molecule very reactive. An antioxidant molecule can supply an electron to neutralize a free radical molecule. A certain level of free radicals is necessary for life because they play an important role in some necessary biological processes. However, a balance of antioxidants and free radicals is also necessary. This balance is disturbed by age that results in a high level of free radicals in your body that may cause certain diseases.

Another very useful carbon nanostructure is a carbon nanotube ➜

Carbon Nanotubes

 **Conceptually:** one graphite layer rolled up into a cylinder makes single walled carbon nanotube (SWCNT).

Coaxial assembly of multiple SWCNT ➜ multiple walled carbon nanotube (MWCNT).

- ◆ **SWCNT**. Diameter 1 nm. Essentially a one dimensional structure; length to diameter ratio exceeds 10,000.
- ◆ **MWCNT**. Typical separation between two neighboring tubes = 0.34 nm; same as that between two layers in a natural graphite.

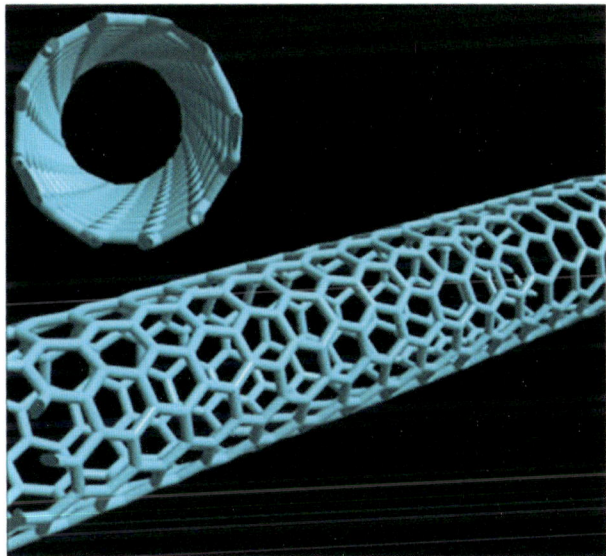

Figure 5.4 An example of single-walled carbon nanotube. The interior of a carbon nanotube is hollow, but can be loaded with a wide variety of molecules.

 *How are carbon nanotubes made?*

Carbon Nanotubes: Fabrication

To make a carbon nanotube, you need the following three typical elements:

♦ A source material that will be used to get the carbon atoms that we need to form the nanotubes.

♦ A collector (or sink) material or object on which the carbon atoms will be deposited to form the nanotubes.

♦ A way for carbon atoms to travel from the source material to the collector (sink) material.

Multiple ways to achieve these three elements => multiple fabrication methods including laser evaporation method.

 *Can you please describe the laser evaporation method to fabricate carbon nanotubes?*

Carbon Nanotube Fabrication: Laser Evaporation Method

Process:

1. A quartz tube contains a graphite target and an inert gas such as argon, which are heated to $1200°C$.

2. The tube also contains laser on one end, and a water cooled carbon collector made of copper on the other.

3. An intense laser beam is incident on the graphite target that evaporates carbon from the target.

4. Argon sweeps the carbon atoms from the high temperature zone toward the water cooled collector.

5. The carbon atoms condense on the cold collector into nanotubes.

By using this method, you can make tubes:

♦ Length: 100 μm

♦ diameter in the range of 10-20 nm

 The nanotube diameters and other properties can be varied by varying the parameters of the process such as

♦ Catalyst composition

♦ Temperature

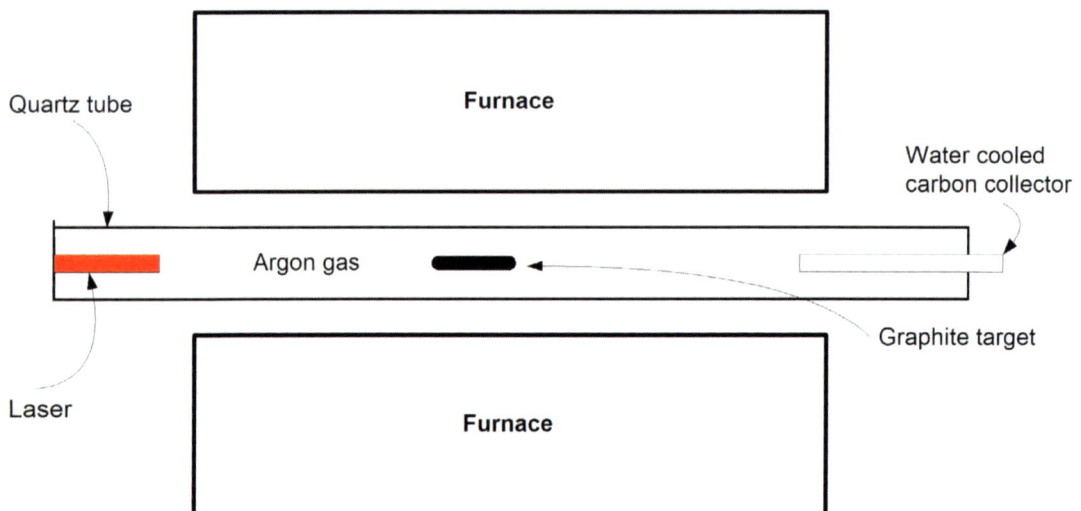

Figure 5.5 Experimental setup for laser evaporation method.

 *What are some of the applications of carbon nanotubes?*

Applications of Nanotubes

- **Conductive property.** Replace transistors as digital switches in chips
- **Strength and high length to diameter ratio.** Use as fibers to make strong materials.
- **Great mechanical properties**. Help make different structures such as combat jackets and space elevators.
- **Strength and flexibility**. Nanoengineering, that is, tools to control or manipulate other nanostructures.

 *Can you elaborate on the strength of carbon nanotubes?*

Strength of Carbon Nanotubes

 **Strength**. The ability of the material to withstand pressure or stress. Carbon nanotubes are the strongest and the stiffest material (or material structure) known.

Definitions Related to Strength:

Pressure. Pressure on a surface is the force applied per unit area in a direction perpendicular to the area:

$$P = F/A; \text{ Units } N/m^2$$

where:

- P is the pressure
- F is the force perpendicular (normal) to a surface
- A is the area of the surface on which the force is being applied.

Stress. Pressure caused by the force applied on an object, say a wire, to stretch or compress it. Same units as the pressure:

$$S = F/A$$

Strain. The deformation caused by the stress on a physical body. For example, in case of a wire, the strain is the amount of stretch as a result of the force applied to stretch the wire:

$$e = \Delta L/L$$

where:

e is the strain, ΔL is the amount of stretch

L is the length of the wire before the force is applied.

Young's modulus. Stress is directly proportional to strain and the proportionality constant is called Young's modulus, Y:

$$S \propto e$$
$$\Rightarrow \quad S = Ye$$
$$\Rightarrow \quad Y = \frac{S}{e}$$
$$\Rightarrow \quad Y = \frac{FL}{A \, \Delta L}$$

- Young's modulus represents the elasticity or the elastic flexibility of a material.
- The larger the value of Young's modulus, the less flexible the material is.
- Young's modulus for steel is about 30,000 times that for rubber.
- Young's moduls for steel is about 0.21 TPa.
- Y for carbon nanotubes is in the range of 1.3 to 1.8 TPa, that is, about 10 times that of steel.

Units of Y:

Y is stress divided by strain. Units of stress are $N/m^2 = Pa$

Strain is dimensionless (no units).

Therefore units of Y are Pa.

Tensile strength. The maximum amount of stress that a material can take before failing.

- *Failure*, an important concept in material science; precise definition depends on the type of the material and the design methodology.

- Example. In case of a wire, the *tensile strength* is the stress required to pull it to the point where it breaks.

- The tensile strength of carbon nanotubes is about 45 billion Pa.

- The tensile strength for high strength steel alloys is about 2 billion Pa. =>

- Carbon nanotubes are about 20 times stronger than steel.

Caution: Distinguish between stiffness and strength.

Thinking of Buckyballs and Nanotubes

Why Balls and Tubes?

- Large surface area ➜ High reactivity (reaction rate)

- Shape is driven by two underlying physical principles:

    1. A physical system tends to be in its equilibrium state, that is, the most stable state.
    2. The most stable state is the state with minimum energy.

- In the nano scale, edge energy matters.

- A tiny piece of graphite would have a lot of atoms at its edge which would lead to instability.

- Given the choice, nano solids would naturally roll themselves up to balls or tubes to get into the stable state by minimizing the total energy

Summary and Conclusions

- Nanofabrication is the fabrication of nanostructures, that is, the material structures with at least one lateral dimension within the nanoscale.

- The history of human development on this planet is largely defined by understanding and using different materials.

- There are two main approaches for nanofabrication: top-down and bottom-up.

- Examples of top-down approach include:
  - Thin-film deposition and growth
  - Nanoimprint
  - Nanolithography
  - Etching technology
  - Scanning probe
  - Mechanical polishing

- Examples of bottom-up approach include:
  - Chemical synthesis
  - Laser trapping

- o Self assembly
- o Colloidal aggregation.

♦ Carbon is the most suitable atom to build nanostructures. Some examples of nanostructures made of carbon atoms are buckyballs and nanotubes.

♦ Buckyballs can be used for medical applications such as delivering drugs to specific targets in an organism.

♦ Due to their special characteristics such as strength, conductivity, and large length to diameter ratio, nanotubes have a wide spectrum of applications in various fields.

Self Test Exercises

1 Calculate the energy of a proton at rest.

2 What is the name of the electromagnetic force that holds buckyballs together?

3 The Young's modulus Y for carbon nanotubes is in the range of 1.3 to 1.8 TPa, that is, about 10 times that of steel. Does that mean that carbon nanotubes are more flexible than steel?

4 A single walled carbon nanotube has a diameter of 1 nm and the length to diameter ratio of 10,000. What is the length of this nanotube in micrometers?

5 A nano planar material has a lot of energy at the edges. Left alone, it will automatically roll itself into what to minimize the energy?

6

Introduction to Nanooptics

Learning Objectives

1. Definitions

2. Nanooptics: Big Picture

3. Nanoscale Photoprocesses

4. Characteristics of Light

5. Nanoscale confinement of Light

6. Nanoscale Confinement of Matter

7. Plasmonics

8. Near-field Optics

9. Photovoltaic cells

10. Optical Lithography

11. Summary and Conclusions

12. Self Test Exercises

Nanooptics

 *Also called Nanophotonics:*

➢ **Optics** is a branch of physics that deals with studying and describing the behavior and properties of light including interaction of light with matter and optical processes, also called photoprocesses.

➢ **Nanooptics** is the optics at nanoscale.

➢ **Categories**. Accordingly, nanooptics has three broad categories:

  o Nanoscale confinement of light
  o Nanoscale confinement of matter
  o Nanoscale optical processes

➢ **Big picture.** I have separated these topics and organized them together in the next illustration for the learning purpose here.

Nanooptics: Big Picture

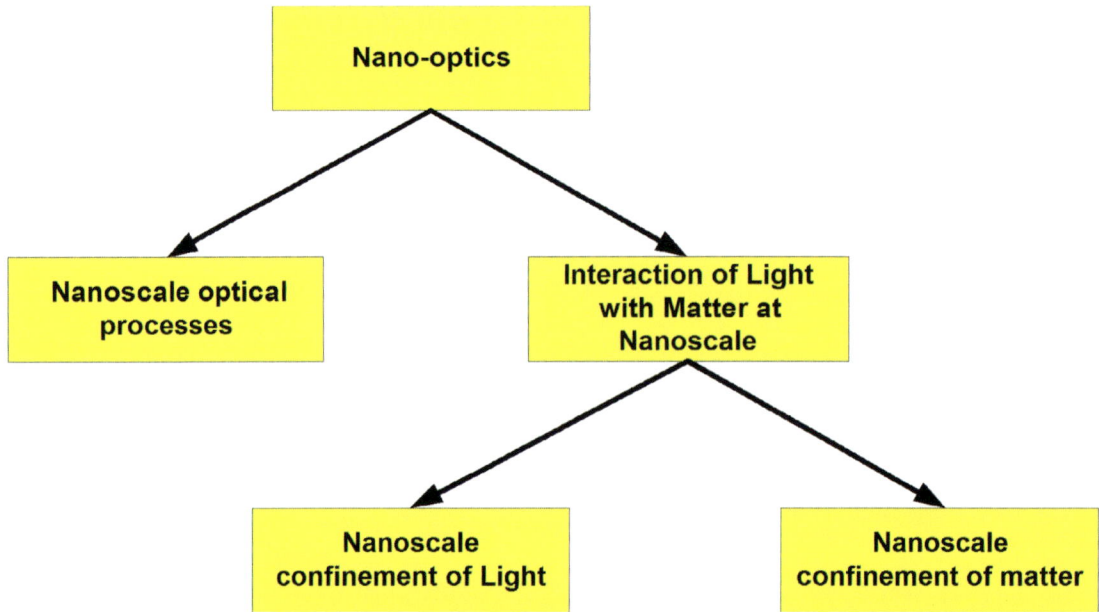

Figure 6.1 Big picture of nanooptics, an illustration.

Optical processes are also called photoprocesses.

We can enter nanooptics through three doorways:

♦ **Nanoscale optical processes**. Phase change of matter induced by light.

♦ **Nanoscale confinement of light**. Confine light to nanoscale in order to induce interaction between light and matter at nanoscale.

♦ **Nanoscale confinement of matter**. Confine matter to nanoscale in order to limit the interaction of light with matter to nanoscale.

Before we dive into the topic, let's review some relevant characteristics of light ➔

Characteristics of Light

♦ Light travels through space (vacuum) with a constant speed, c, given by:

$$c=3\times0^8 \ m/sec$$

♦ The wavelength of light through space (vacuum) can be calculated by the following formula:

$$f=c/\lambda$$

Example:

Ultraviolet light is in the frequency range approximately from 7.5×10^{14} to 3×10^{17} Hz. The higher end of this range corresponds to:

$$\lambda=c/f$$
$$= \ (3\times10^8 \ m/sec)/(3\times10^{17} \ /sec)$$
$$= \ 1\times10^{-9}m=1 \ nm$$

Another example:

X rays are electromagnetic waves in the frequency range of 3.0×10^{17} to 3×10^{19} Hz.

The higher end of the X rays frequency range corresponds to:

$$\lambda=c/f$$
$$=(3\times10^8 \ m/sec)/(3\times10^{19} \ /sec)$$
$$=0.01\times10^{-9}m=0.01 \ nm$$

So, X-rays and higher end of UV light may be considered as composed of nanoparticles: photons.

Nanoscale Photoprocesses

 **Nanoscale photoprocesses**: The photoprocesses that can be confined to well defined regions of nanosize.

♦ This facilitates fabricating structures in a precise arrangement and geometry.

♦ These processes can be used in nanolithography and to make nanosensors and nanoscale optical memory.

In order to perform a nanoscale photoprocess, you need to confine light to a nanoscale region.

 *How do you confine light to a nanoscale region?*

Nanoscale Confinement of Light

◆ Several ways to confine light to a nanoscale

◆ One of these methods is called *near-field* optical propagation.

◆ **Example**: light emanating through a tip opening much smaller in size than the wavelength of the light, e.g., by using a metal-coated and tapered optical fiber.

◆ **A principle of confinement:** Total internal reflection of light.

Assume a ray of light strikes a medium boundary at an angle larger than a certain angle called *critical angle* with respect to the normal to the surface. If the refractive index is lower on the other side of the boundary no light can pass through, so effectively all of the light is reflected.

The *critical angle* is defined as the angle of incidence above which the total internal reflection occurs.

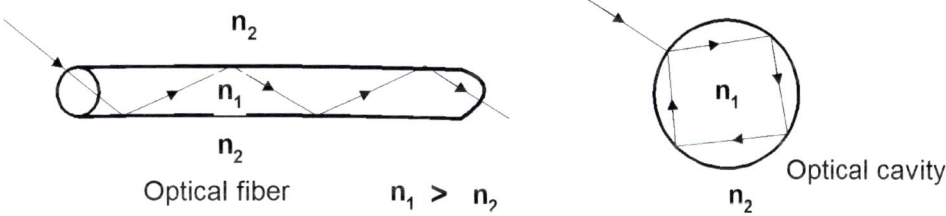

Figure 6.2 Confinement of light.

Photon (light) is a particle that has no mass. Other particles such as electrons have mass. If massless particles such as photons can be confined, how about confining the matter?

 *So, what is nanoscale confinement of matter?*

Nanoscale Confinement of Matter

Main Points:

♦ Matter confined to a nanoscale is called nanoparticles or nanostructures.

♦ In nanooptics, nanoscale confinement of matter deals with making nanomaterials for optics.

♦ To make the optical nanostructures, the dimensions of the matter can be confined by using various methods.

♦ Nanoparticles with unique optical properties are already being used in products and applications such as ultraviolet absorbers in sunscreen lotions.

♦ Nanoparticles in optics may be inorganic or organic.

♦ Examples of inorganic nanoparticles: the metallic nanoparticles that exhibit enhanced electromagnetic field and unique optical response; constitute an area called plasmonics.

♦ An example of organic nanoparticles is nanomer, which is a nanosize oligomer.

♦ An oligomer is a small polymer molecule, that is, a polymer with a small number of structural repeat units called monomers. For example a sort, single, stranded DNA fragment can be called an oligomer.

 *What are some of the basic concepts involved in the confinement of matter?*

Confinement of Matter

 Some Basic Concepts

- **Band.** The range of energy that an electron in a solid state material is allowed to have or not have. If this is the energy range that an electron cannot have, it's called *forbidden band*.

- **Valence band.** The highest energy band that the electrons in a solid can have.

- **Conduction band.** The range of energy, higher than the valence band, that when an electron has, it is free to move in the solid material under the influence of external electric field, that is, the material can conduct electricity.

- **Band gap.** The energy gap between the valence band and the conduction band of a solid state material. This is the gap in which an electron cannot exist. In other words, the electrons in the valence band must cross the band gap to make the material electrically conductive.

- **Semiconductor.** A material that has electrical conductivity between that of a conductor and an insulator. Insulators have larger band gap than that of a semiconductor.

 *The electrical properties of semiconductors can be controlled by controlling the band gaps.*

One of the important concepts in nanotechnology is quantum confinement.

 *What is quantum confinement?*

Confinement of Matter
Quantum Confinement

 **Quantum confinement.** Confinement of matter controlled (or explained) by quantum physics.

Various quantum confined structures developed from semiconductors by manipulating the band gaps:

➢ **Quantum well.** This results from confinement in one dimension. A thin layer of a semiconductor with a small bandgap is sandwiched between two layers of a semiconductor with wider bandgaps. The heterojunction between semiconductors with small bandgap and large bandgap establishes what is called a potential well (or quantum well) that confines the electrons in the material with smaller bandgap.

➢ **Quantum wire.** This results from confinement in two dimensions. Two dimensional confinement ➔ The electrons in the quantum wire are only free to move in one dimension, that is, along the length of the wire.

➢ **Quantum dot.** This results from confinement in three dimensions. Therefore an electron in a quantum dot is confined in a three dimensional box with the dimensions ranging from one to tens of nanometers.

 *What are the effects of these dimensional confinements?*

Effects of Dimensional Confinement

The dimensional confinement has profound effect on the optical (and electrical) properties of materials.

- ♦ The optical properties of quantum confined nanostructures depend on the size (e.g. length) of confinement; this dependence can be exploited.

- ♦ The bandgap decreases with the increase in the dimensions of confinement.

- ♦ The quantum confined structures are remarkably efficient lasing media.

- ♦ The optical (and electronic) properties of quantum confined structure exhibit strong dependence on the *dielectric constant* of the surrounding media; exploit this dependence for manipulating properties for different purposes by tuning the *dielectric constant*.

Example: *Lasers supported by quantum wells hold a lion's share of the solid-state laser market.*

What is dielectric constant?

Dielectric Constant

 **Dielectric constant**. The capacity of a material (or medium) to retain charge.

Consider a capacitor with charge $+Q$ on one plate and $-Q$ on the other. The capacitance of the capacitor:

$$C = Q/V$$

V is the potential difference between the two plates.

Assume:

C_{med} is the capacitance of this capacitor when the plates are separated by the medium

C_v is the capacitance when the plates are separated by the vacuum.

In this case, dielectric constant of the medium is given by:

$$K = C_{med}/C_v$$

Another important concept in nanotechnology is: Plasmonics ➔

Plasmonics

The study of metallic nanostructures and their applications

Why is it called plasmonics?

 **Plasma.** Ionized bulk matter such as gas, that contains free electrons which are not bound to any atom, and therefore it also contains positively charged ions.

A distinguished feature of plasma

A charged particle in plasma influences several other charged particles nearby. These influences (changes or oscillations) in the plasma gives rise to waves called plasma waves.

 **Plasmon.** Just like photon is a quanta (particle) of light (or EM waves), plasmon is a quanta of plasma oscillations (or waves).

Examples:

- ◆ If some electrons are displaced slightly, the electric force from the positively charged ions will pull them back and start oscillations and hence waves.

- ◆ The changing density of electric charges at a given point will give rise to waves.

The quanta of these waves are called plasmons.

 *Are there any applications of plasmonics?*

Applications of Plasmonics or Metallic Nanostructures

Applications are based on three features of metallic nanostructures:

- ✓ Local field enhancement
- ✓ Wave excitation
- ✓ Dielectric sensitivity

Local field enhancement. The EM field around the surface of a metallic nanostructure is enhanced:

- ➤ The field enhancement could be large enough to see the Raman spectra of a single molecule.
- ➤ This confinement of electromagnetic radiation around the surface has been manipulated in photofabrication, a method called plasmonic printing.
- ➤ It has also been used for apertureless near-field microscopy.

- ➤ **Evanescent wave excitations**. The evanescent wave caused by the enhanced EM field near the surface of a metallic nanostructure can be used to excite an optical transition in fluorescent molecules in a preferential manner

  - ○ This has found its use in various fluorescence-based optical sensors, and o Other optical signal related applications.

➢ **Dielectric sensitivity.** The frequency of the plasma waves at the surface of a metallic nanostructure is sensitive to the dielectric environment. This sensitivity is used to detect the biological analytes, which bind to the surface of the nanostructure.

An analyte is a substance or a chemical constituent whose properties such as amount are being determined in an analysis. For example, glucose is an analyte in a blood test being run to determine the amount of glucose in the blood.

This has been confinement of matter.

Back to confinement of light which can be accomplished by using near-field optics.

What is near-field optics?

Near-Field Optics

- **Near field.** The electromagnetic radiation (light) within the distance of one wavelength magnitude (or a small number of wavelengths) from the emitter such as antenna or aperture is called near field. This region is called near field region.

- **Far field.** The EM radiation beyond the near field region is called far field, and the region of the far field is called the far field region.

- The transmitted EM radiation exhibits different behaviors in near field region from that in far-field region:

  - The radiation power (the field strength) decreases as a cube of the distance from the emitter in the near field, while it decreases only as a square of the distance in the far field.

  - The diffraction pattern of light in the near field region differs significantly from that in the far field region.

  - The radiation in the near field can have a greater resolution than the radiation in the far-field because the diffraction effect has not taken over yet.

Can near-field optics be used to improve optical microscopy?

Near-field Optics: Microscopy

Poor resolution limit on traditional optical microscopes comes from the fact that they are based on far-field optics.

Solution: near-field optics microscopy.

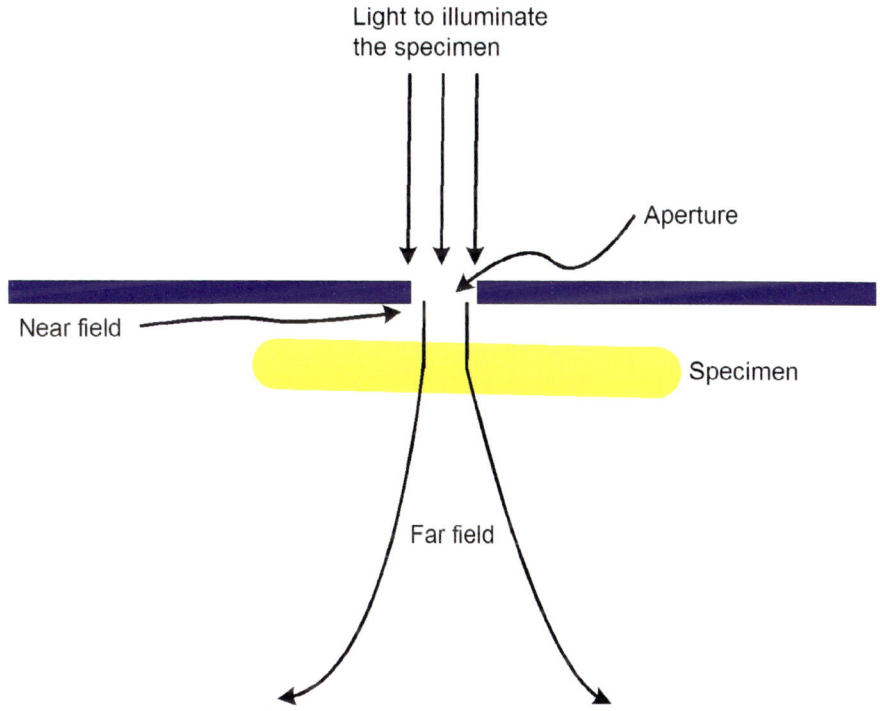

Figure 6.3 Illustration of near-field microscope.

- ◆ Specimen in the near-field ➔ better resolution
- ◆ The specimen illuminated through a sub-wavelength sized aperture.
- ◆ A spatial resolution of about 20 nm can be achieved.

Are there other applications of near-field optics?

Near-field Optics

More Applications

Trapping particles. A component of the near field (near the emitting material) is a standing wave called evanescent wave which decays exponentially with the distance. This near field wave can be used to exert optical *radiation pressure* on nano particles to slow them down and trap them.

Optical storage. A new storage technology based on near field optics, called near field optical storage is being developed that can store up to 150 Gbytes of data on cheap optical media similar to DVDs. Near field optical recording requires a gap of the order of 25 nm between the read/write head and the media.

Radiation pressure is the pressure exerted by EM radiation on a surface or object that it hits. Remember that EM radiation (light) has energy and therefore momentum.

In optics, nanotechnology can also help make better photovoltaic systems or photovoltaic cells.

What are photovoltaic cells?

Photovoltaic Cells

The word **Photovoltaic** has a Greek origin: *phos* means light, and "voltaic", means electrical.

Photovoltaic Cell: A device that converts light into electricity.

Solar cell: A type of photovoltaic cell; converts Sun light into electricity.

Photovoltaic effect. The phenomenon in which matter emits electrons after absorbing photons (light).

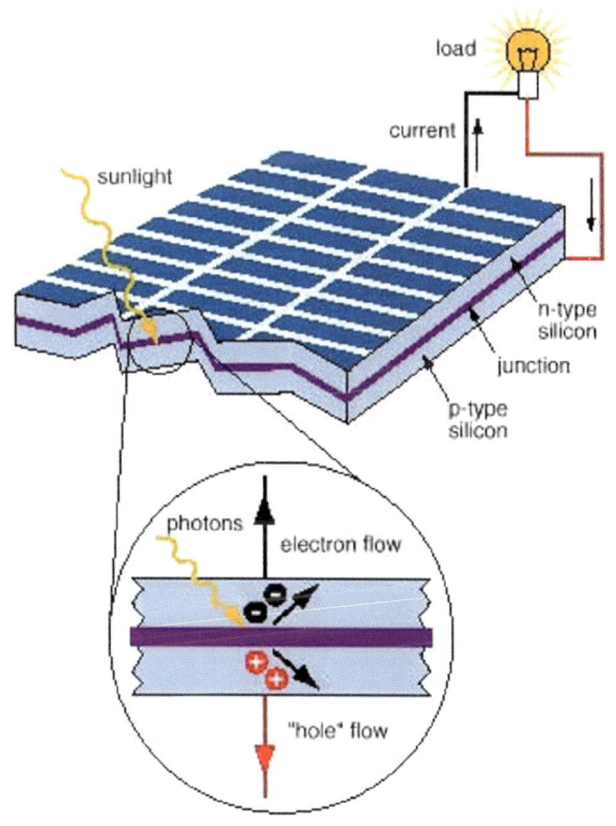

Figure 6.4 Illustration of a photovoltaic cell.

 *So, what is the physics behind the photovoltaic effect?*

Physics Behind Photovoltaic Effect

Following is the physics behind the photovoltaic process:

1. A photon hits the solar panel.

2. The photon is absorbed by the material, e.g. a semiconductor material such as silicon, in the panel.

3. The energy of the photon knocks an electron off an atom to make it free to flow through the material; the atom is ionized.

4. The free electrons flow to produce electricity: direct current (DC).

5. The direct current, if desired, can be converted into alternating current (AC) by using appropriate devices such as inverters.

6. The AC power enters the utility panel from where it is distributed.

 *What does this has to do with nanotechnology?*

Well, it depends on how you manipulate photovoltaic effect create energy ➜

Three Generations of photovoltaic cells

First generation. Silicon-wafer based solar cells. Over three decades old.

Second generation. Thin-film solar cells. About a decade old.

Third generation. Do not rely on P-N junction. Example: Nano/polymer PV systems. New.

Can you say more about the first generation photovoltaic cells?

1ˢᵗ Generation Photovoltaics

Si wafer based solar cell

- Consist of a large area
- Single layer of p-n junction diode
- Currently 90% Market Share, over 30 years experience

Problems:

- Silicon not a good absorber of light ➔ very thick cell
- Wafers are fragile ➔ challenge in handling ➔ complicates processing
- High material cost, poor efficiency

 *How about 2ⁿᵈ Generation?*

2nd *Generation Photovoltaics:*
Thin Film PV System

- Use of thin film of semiconductor ➔ Reduce the amount of light absorbing material ➔ Reduce processing cost.
- The thickness of thin films cound vary from a fraction of a *nanometer* to a few micrometers.
- Multi-thin films have efficiency higher than that of silicon-wafer.
- Material: Amorphous or micro-crystalline Si, CIS, Cadmium Tellurium, CIGS)
- ~6% Market share and <5 years experience; growing
- Low cost and light weight.

Concern/Issue: low efficiency, at least for now.

   *How do you define the efficiency of a photovoltaic system?*

Efficiency of PV Systems

♦ The main factors driving the adaptability of the photovoltaic systems are the cost and efficiency.

♦ Efficiency of a solar device is defined as the ratio of the maximum power point produced by the device to the input power (radiation absorbed per unit time) to the device:

$$\eta = P_{out}/P_{in}$$

Exercise. A solar roof is absorbing the radiation from the Sun at the rate of 1300 W/m^2. A solar cell on that roof has a size of 105 cm^2 and produces 1.6 W of power. What is the efficiency of the solar cell?

Solution:

```
Pout = 1.6 W / 105 cm²

Pin = 1300 W/m²

    η=Pout/Pin

     =(1.6 W/105 cm²)/(1300 W/m²)

     = 152/1300

     = 0.117

     = 11.7 %

So, the efficiency is 0.117 ➔ 11.7%.
```

Now, how does nanotechnology come into the picture?

3rd Generation Photovoltaics
Nano / Polymer systems

- No longer relying on p-n junction to separate carries.
- Compositing polymer and nano particles together to form thin multi-spectrum layers to improve the efficiency.
- Should be cheaper and easier to manufacture.

Photovoltaic cells are on example of how nanotechnology is playing an important role in a rapidly growing alternative energy field called clean technology or cleantech.

Another application of nanooptics: Optical Lithography ➜

Optical Lithography

Optical lithography, also called **photolithography**, is a lithographic technique that uses light to transfer pattern from a photomask to the light sensitive material on the substrate called resist.

Why optical lithography?

It can provide resolution in the nanoscale.

Process

Following is the typical process used in optical lithography:

1. A silicon wafer is spin-coated with a light-sensitive resist.
2. Light of suitable wavelength such as ultraviolet light is projected through the pattern on the photomask on to the layer of the resist.
3. The regions of the resist exposed to light become more soluble. This way the pattern (from the photomask) is reproduced on the resist layer.
4. The semiconductor material underneath the resist that has been exposed by light through the holes in mask is etched away by using an acid.
5. A solvent is used to wash away the resist from the wafer.
6. The pattern from the photomask has been transferred to the semiconducting material, e.g., the silicon wafer.

Summary and Conclusions

- ➤ Nanooptical effects can be achieved by three methods:

  - ▪ **Nanoscale optical processes**. Phase change of matter induced by light.

  - ▪ **Nanoscale confinement of light**. Confine light to nanoscale in order to induce interaction between light and matter at nanoscale.

  - ▪ **Nanoscale confinement of matter**. Confine matter to nanoscale in order to limit the interaction of light with matter to nanoscale.

- ➤ The optical (and electronic) properties of quantum confined structures exhibit strong dependence on the dielectric constant of the surrounding media. Therefore, we can exploit this dependence to manipulating material properties for different purposes by tuning the dielectric constant.

- ➤ Metallic nanostructures can be used to create plasma waves, which have several properties including photofabrication and apertureless near-field microscopy.

➢ In addition to improving optical microscopy, near-filed optics has many other applications including particle trapping and optical storage.

➢ Other applications of nanooptics include 3rd generation photovoltaic cells and optical lithography.

Self Test Exercises

1 Matter confined to a nanoscale makes what kind of particles?

2 The electrical properties of semiconductors can be controlled by what characteristic of semiconductors?

3 In a quantum well, particle is free to move in how many spatial dimensions?

4 The optical and electronic properties of quantum confined structures exhibit a strong dependence on the dielectric constant of the surrounding media. Why is that important?

5 Rearrange the entries in the second column of the following table to match them correctly with the entries in the first column.

Entity	Definition
Plasmon	Quanta of light
Photon	Quanta of oscillations in plasma
Plasma	Ionized bulk matter that contains free electrons
Plasma waves	Changing density of electric charges at a given point

7

Introduction to Nanoelectronics

Learning Objectives

1. Definitions

2. Current Status of Nanoelectronics

3. Background: Microelectronics

4. Moore's Law

5. Electron Beam Lithography

6. Nanocircuit, Nanoradio, and Nanowires

7. Methods To Fabricate Nanowires

8. Molecular Electronics

9. From Binary Computing to Quantum Computing

10. NEMs

11. Summary and Conclusions

12. Self Test Exercises

Nanoelectronics

 **Nanoelectronics**. Branch of nanotechnology that deals with electronic components and devices.

Main Goal. To store, transmit, and process information by taking advantage of the properties of matter at nanoscale that are distinctly different from the properties at macro scale.

Components. Nanoelectronics is based upon nanoscale components or their ordered assemblies:

- Molecular electronics
- Nanotubes
- Nanowires
- NEMs

Fabrication Methods:

- Nanolithography
- Vapor deposit methods
- Other fabrication techniques

Nanoelectronics: Current Status

Currently two active fields of research:

1. The characterization and fabrication of individual components that could replace the macroscopic/microscopic silicon components with nanoscale systems.

   Examples:

   - Molecular diodes

   - Single atom switches

   - Better control and understanding of the transport of electrons in quantum dot structures.

2. Investigation of potential interconnects.

   Example:

   Investigate carbon nanotubes and self-assembled metallic and organic structures as potential electronic components.

First, there was microelectronics

 *What is microelectronics?*

Microelectronics: Concepts

 *Definitions*

- **Electronics.** The branch of physics that deals with the flow of charge through various materials and devices such as semiconductors and vacuum tubes.

- **Microelectronics.** The branch of electronics that deals with very small electronic components and devices, which are generally made from semiconductors by using techniques such as lithography. Microelectronics is largely based on microchip.

- **Microchip.** An electronic device that consists of an integrated circuit embedded on a semiconducting material. There are many kinds of chips such as central processing units and memory chips.

- **Integrated circuit.** An electronic circuit made of tiny components called transistors.

- **Transistor.** A semiconductor device that can act as an amplifier or a switch in an electric circuit. It's a basic building block of the circuitry used in modern electronic machines such as computers and cellular phones.

- **Semiconductor.** A material that has the electrical conductivity halfway between that of a conductor such as a metal and an insulator such as rubber.

The history of microelectronics can be described in terms of Moor's Law.

 *What is Moore's Law?*

A Law of Microelectronics: Moore's Law

 **Moore's Law**. The number of transistors on a chip doubles about every two years.

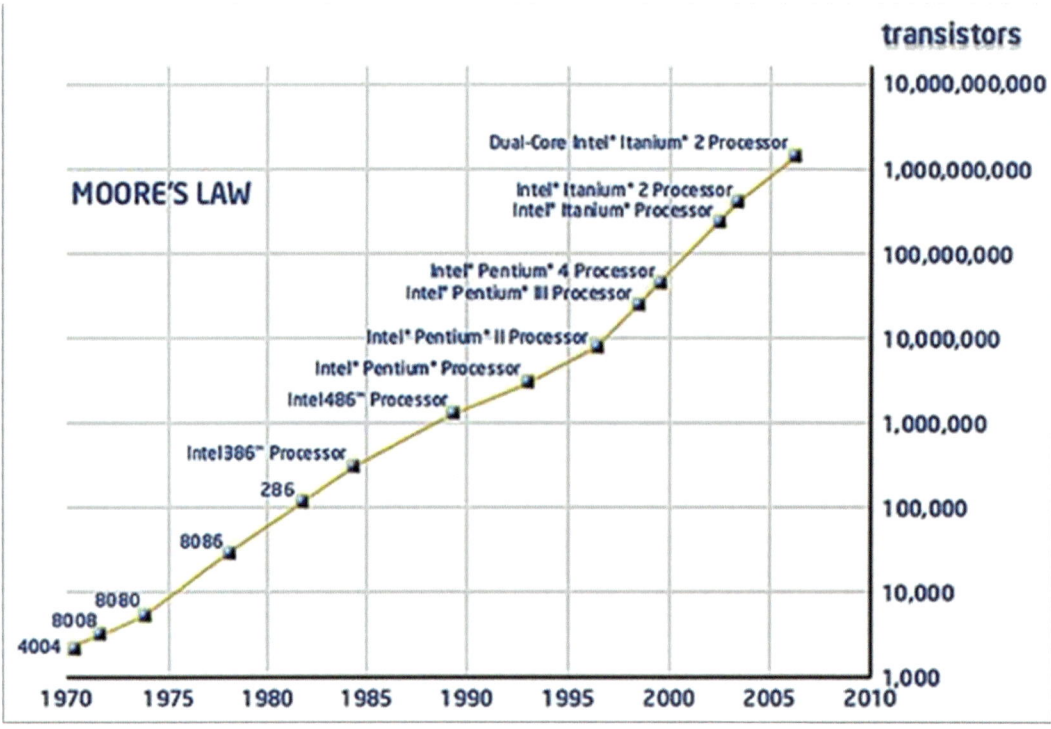

Figure 7-1. Illustration of Moore's Law.

 *So what?*

Implications of Moore's Law

There are a number of implications of this law including the following:

- ◆ **Power**. As the number of transistors on a chip is increased, the power or performance of the chip (or the microprocessor) will also increase.

- ◆ **Cost**. As the number of transistor on a chip increases, the cost per transistor reduces. That means more power (performance) with less money.

- ◆ **Size**. In order for the increased number of transistors to fit on the same chip, their size has to be decreased. In general, the individual feature sizes of electronic components decrease as a result of increase in number of transistors.

About the size:

Before we reach nano size, the rules of physics change ➔ discontinuity in Moore's law

 *Can Nanotechnology save Moore's Law or even do better?*

Yes,

For example, electron microscope (a nano tool) can help in Lithography ➔

Electron Beam Lithography

The basic process to generate a pattern:

1. The electron gun of a scanning electron microscope emits electrons.

2. The beam of electrons is focused on the surface of the resist-coated substrate such as a semiconductor wafer.

3. The computer-controlled pattern generator moves the electron beam across the surface to generate the integrated circuit (IC) pattern.

The effective resolution depends on

- The wavelength of the electrons

- The diameter of the electron beam

- The material used as a resist.

 *A line width of 10nm or smaller on the pattern can be achieved by using electron lithography. The electron beams with diameter of the order of nanometers can be produced.*

Limits to the resolution are posed by:

- Scattering in the resist material, which makes the resolution worse than the beam width.

- The secondary electrons, that is, the electrons generated as a result of ionization.

 *The scattering effect can be reduced by using high energy electrons and by using thin resist.*

Nanocircuit

Three elements:

1. Transistors
2. Interconnectors
3. Architecture

All three elements are dealt within the nanoscale.

Carbon nanotubes can help build nanocircuits.

 *What are carbon nanotubes?*

Carbon Nanotubes

Carbon nanotube. A cylindrical nanostructure with diameter in the nanoscale. These structures can have length to diameter ratio of greater than 40 million ➔ ideal for using as nanowires.

Properties of nanotubes that make them ideal candidates for making nanocircuits:

♦ Nanotubes are smaller than any traditional wires being used in the electronics industry.

♦ They have excellent material strength.

♦ They can tolerate extreme temperatures.

♦ They are non-reactive.

♦ Current can pass through them with almost zero resistance.

Carbon nanotubes have several applications in Nanoelectronics; can be used as:

♦ Conductors
♦ Semi-conductors
♦ Transistors
♦ Electronic devices such as *nanoradios*

Challenge: Large scale production of chips based on carbon nanotubes.

 *What is a nanoradio?*

Nanoradio

 **Nanoradio** is a radio receiver or transmitter constructed on a nanoscale.

October 2007. A team of physicists at University of California, Berkeley reported the first development of a radio receiver from a single carbon nanotube.

The single nanotube offers the functionalities of all major components of a radio:

♦ Antenna

♦ Amplifier

♦ Demodulator

♦ Tuner.

 *The signals are received via the high frequency mechanical vibrations of the nanotube as opposed to the traditional electrical means.*

 *How does a nanoradio work?*

Nanoradio: How does it Work?

Here is how it works:

1. A carbon nanotube, typically 10 nanometers in diameter and several hundred nanometers in length, is contained in a vacuum.
2. One end of the nanotube is connected to an electrode of a battery, say *first electrode*.
3. The *second electrode* is placed a short distance from the nanotube's other end.
4. The tube charged this way will vibrate in tune with any external electromagnetic signal, and therefore effectively acts as an antenna.
5. The vibration frequency can be adjusted to a desired value by changing the applied voltage. This allows to tune the radio to different carrier frequencies.

The *field emission effect* causes a current flow as the emitted electrons tunnel across the gap between the nanotube and the second electrode. This current represents an amplified version of the radio signal; no demodulation is necessary.

Field emission refers to the emission of electrons from the surface of a material due to the presence of external electric field. The current produced by the emitted electrons is directly proportional to the applied field.

In addition to carbon nanotubes, nanowires can also be used to put together a nanochip ➔

Nanowires

Nanowires. Wires that have thickness of the order of nanometers and unconstrained length.

Different kinds of nanowires including the following:

♦ Metallic (or conducting) nanowires made of materials such as Ni, Pt, and Au.

♦ Semiconducting nanowires made of material such as Si, InP, and GaN.

♦ Insulating nanowires made of materials such as SiO_2 and TiO_2.

♦ Molecular nanowire made as a chain of molecular units such as DNAs.

Three common methods to fabricate nanowires:

Suspension method. A wire is held in a vacuum chamber at two ends. A bigger wire suspended this way can be chemically etched or bombarded with high energy particles to turn it into a nanowire.

Deposition method. The nanowire is deposited on a surface of different material or nature, for example, a strip of metallic atoms deposited on a non-conducting surface.

Synthesis method. The repeating elements are joined (or synthesized) together to make a nanowire such as a molecular nanowire.

A popular technique for creating a semiconductor nanowire is the Vapor-Liquid-Solid (VLS) synthesis method ➔

Vapor-Liquid-Solid (VLS) Synthesis Method

The source material. Laser ablated particles or a feed gas such as silane ($SiH4$).

Catalyst. The source is exposed to a catalyst.

For nanowires, liquid metals such as gold nanoclusters make the best catalysts.

Some chemical reactions involved in the synthesis process:

$$SiH_4 \rightarrow Si(nanowires) + 2H_2$$

$$SiCl_4 + 2H_2 \rightarrow Si(nanowires) + 4HCl$$

The advantages of this method include the following:

♦ The synthesis requires relatively low growth temperature, that is, less than $600°C$.

♦ It's easy to control the diameter of the nanowires.

♦ The wires have good crystalinity and less defects.

The main disadvantage:

The method uses a catalyst and therefore contamination from catalyst is the main concern.

The solution to this problem is the thermal evaporation method ➔

Thermal Evaporation Method

- ◆ **Source**. Uses solid precursor (or source) .

- ◆ **Reaction.** Involves the following reaction:

  $$2SiO \rightarrow Si(nanowires) + SiO_2$$

Advantages:

- ◆ There is no contamination from the catalysts.

- ◆ No toxic gas is involved.

Disadvantages:

- ◆ High growth temperature is required (> 1000°C).

- ◆ Diameter of the nanowires is difficult to control.

- ◆ There is more probability for defects.

 *Remind me, what are the uses of nanowires?*

Uses of Nanowires

Still at research level:

- Create *n-type* and *p-type* semiconductors by chemically doping a semiconductor nanowire.
- Create p-n junctions: simplest electronic device.
- Connect several P-N junctions together to build logical gates: AND, OR, NOT.
- Conducting nanowires can be used to connect molecular-scale entities in molecular computing.
- Use as:
    - Transistors.
    - Field emitters
    - Leads in biosensors

 *What is molecular electronics?*

Molecular Electronics

 **Molecular electronics**. The branch of electronics that deals with using molecules as building blocks to fabricate electronic components:

- ✓ Active components such as transistors
- ✓ Passive components such as resistive wires

Molecular electronics provides control to the level of molecules ➔ the size reduction of electronic components

The control to the level of molecules is made possible by nanotechnology tools such as:

- ♦ Scanning tunneling microscope (STM)
- ♦ Atomic force microscope (AFM)

A switch is the heart of digital computing/electronics, for example :

> 1: on
>
> 0: off

Talking about molecular electronics, it's natural to think of molecular switch.

 *What is a molecular switch?*

Molecular Switch

Molecular switch. A molecule (or an array of molecules) that can be shifted back and forth between two or more stable states.

❖ **The shift can be caused by the change in some suitable variable such as**:

- ♦ Current
- ♦ pH
- ♦ Temperature
- ♦ Light

❖ **In General, Molecules can be used as building blocks in electronic devices including chips, and can be used in various applications and devices including:**

- ♦ Transistors
- ♦ Interconnects such as wires
- ♦ Assemblers
- ♦ Memories
- ♦ Insulators

So far, computing has been based on the binary concept: true or false, on or off, yes or no and so on…

Can you please explain computing in terms of the binary concept?

Binary Computing

Current computing is based on *binary* algebra.

❖ **In other words**:

Digital electronics and computer chips use the logic of binary (also called boolean) algebra, which is based on two digits: 1 and 0.

Basic terms used in binary algebra:

- ◆ **Bit**. A binary digit which can take one of two possible values 1 or 0.

- ◆ **Binary number**. A number written in the binary numeral system, that is, a sequence of 1s and 0s.

- ◆ **Binary numeral system**. A base-2 number system in which numbers are written in terms of bits, that is, 1 and 0.

To understand binary system, recall decimal system:

❖ **Decimal Numeral System**

A base-10 number system.

Example. Three hundred and fifty seven will be written as:

$$357$$

Equivalent to:

$$3*10^2 + 5*10^1 + 7*10^0 = 300 + 50 + 7 = 357$$

(The symbol * represents multiplication)

The general process to obtain the value from a number in decimal system written in a combination of digits is:

1. Find the value of each digit in the number, which is given by:

    c_i*10^i

    where i is the position of the digit in the number counted from the right with the position of the first digit being 0. For example, the value of i is 0 for the digit 7, 1 for 5, and 2 for 3 in the number 357; c_i is the digit itself.

2. Add the values of all the digits in the number.

Example:

The number 1037 can be written as:

$$1*10^3 + 0*10^2 + 3*10^1 + 7*10^0 = 1037$$

Because we know that 10^0 is equal to 1 and anything multiplied by 0 is 0, we could have simply written it as:

$$1*10^3 + 3*10^1 + 7 = 1037$$

Or

$$10^3 + 3*10^1 + 7 = 1037$$

because multiplying by 1 does not change anything.

Now, consider an example of a number written in binary numeral system:

    1111111

Because it has base 2 instead of base 1, the value of each digit (bit in this case) will be determined by the following formula:

c_i*2^i

The value of C_i is either 1 or 0. If it's 0, we can simply ignore that digit in the sum because it's contribution to the sum will be zero.

Following this, the value of 1111111 is:

$1*2^7+ 1*2^6 + 1*2^5+ 1*2^4+ 1*10^3+ 1*2^2+ 1*2^{1} + 1*2^0$

$= 128 + 64 + 32 + 16 + 8 + 4 + 2 + 1$

$= 255$

How many numbers can we represent with a given number of bits?

Answer: As many unique combinations as you can make of the given number of bits. Each unique combination will represent a unique number.

A total of n bits can represent 2^n unique numbers.

After this review, we are ready for exploring the concept of quantum computing →

Quantum Computing

Quantum computer: A computer that uses quantum properties of particles to represent, store, and process data.

Quantum computing. Computing in which data is represented by quantum bits, called qubits, as apposed to bits in conventional (or binary) computing.

The Idea Proposed in 1982 by the famous physicist and a noble laureate Richard Feynman.

Involves Controlling and manipulating atoms at individual levels.

Speed. A quantum computer, once built, expected to be exponentially faster than a classical (that is binary) computer. This is made possible by a quantum process called *quantum parallelism*.

 *What is quantum parallelism?*

It is a quantum mechanical process based on the superposition of base states of a particle which can be exploited to perform calculations in parallel.

An example ➜

Quantum Parallelism in Form of Qubits

♦ Data (or memory) in classical (conventional) computers is made of binary bits: each bit can hold a value of 1 or 0.

♦ Data in a quantum computer is made of what are called *qubits*, where a qubit can hold one of many (possibly infinite) values at the same time.

♦ The qubit may represent any observable quantity that may have at least two final discrete values and is conserved over time.

♦ **Example**. An electron may have a spin of +1/2 or −1/2 called *up* or *down* respectively.

♦ **Parallelism**. Continuing with our example, the state of electron at a given time is a combination of up or down with some probabilities attached to each of these two discrete values, and can be represented by:

$$|\psi> = a|up> + b|down>$$

Where:

✓ a^2 is the probability that the electron has spin +1/2 .

✓ b^2 is the probability that it has spin −1/2

The coefficients a and b can have infinite number of values ➔ There are infinite number of possible states of the electron ➔ Infinite number of possible values for the *qubit* that represents the spin of the electron.

 *So, how come we are not running around with quantum computers in our hands? What are the challenges of quantum computing?*

Challenges of Quantum Computing

Include:

- ♦ **New techniques**. Need to discover new techniques compatible with quantum computing. For example, if the quantum computers were to be used to facilitate measurements, the measurement techniques must lend themselves to quantum computing.

- ♦ **New algorithms**. New algorithms needed for solving problems that will take advantage of quantum processes and quantum computing.

- ♦ **Decoherence**. Qubits are superpositions of all the possible states of a system ➔ Qubits are correlated with each other ➔ If one qubit is corrupted by some external phenomenon such as passing cosmic rays, the whole computation will become useless. This situation is called de-coherence because all the qubits are not coherent with each other after something wrong happens to one qubit.

 *Quantum decoherence is defined as the mechanism in which different states of a system seem to collapse as a result of their interaction with the external environment.*

NEMS and Nanomotors are other examples of Nanoelectronics devices ➔

NEMS

Nanoelectromechanical System (NEMS). Nanoscopic system that has characteristic length of less than 100 nm, and combines the electrical and mechanical components.

Before NEMS, there were MEMS:

Microelectromechanical system (MEMS). An integration of mechanical structures (e.g. moving parts) with micro electronics.

- Mechanical actions of MEMs are usually controlled by a computer.

- These systems are called micro systems because they may have components as small as a few micrometers and can function on the microscale.

- MEMs have already been used as pressure sensors, and in monitoring car engines and air conditioning systems.

- NEMs are the small integrated systems that can function at nanoscale and combine electrical and mechanical components

The whole MEMS or NEMS does not have to be of micro or nano size. If the functional components of the system are of micro/nano size, it may be referred to as MEM/NEM.

Can you please cite an example of NEMS?

Nanomotor

General idea. Convert energy such as electrical energy or chemical energy into movement (work) corresponding to force of the order of piconewton: 10^{-12} N.

- ✓ A water droplet is equivalent to about 10 micronewton, i.e., 10 million piconewton.

- ✓ An eylash can exert force of about 100 nanonewton, that is, 100, 000 piconewton.

Figure 7-2 Computer generated image of a nanomotor invented by a research group at UCB.

A nanomotor invented by a research group at UCB (Zattl Research Group) has the following characteristics:

- ◆ The nanomotor is about 500 nm across.

- ◆ The motor is capable of converting one form of energy into another at the force scale of piconewtons.

♦ The motor's shaft is a multi-walled carbon nanotube, which is welded both to the rotor and the fixed anchor.

♦ The gold rotor, nanotube anchors, and opposing stators were patterned around the chosen nanotubes using electron beam lithography.

♦ The gold rotor turned on the carbon nanotube shaft, powered by two charged stators patterned on a silicon surface.

NEMS Construction and Uses

Construction:

Top down approach. NEMs are produced by nanomachining.

Bottom up approach. Production uses biochemical processes.

Uses:

♦ Nanosystems help in characterization and fabrication of nanosystems. **Example**: Microcentilever with sharp nanotips used in STM and AFM.

♦ NEMS field also provides experimental and computational tools.

♦ Carbon nanotubes and nanotube-based products such as electrodes and sensors

♦ Biological motors such as the ones based on DNA

♦ Molecular gears made by attaching benzene molecules to the outer walls of carbon nanotubes

♦ Quantum wires

♦ Quantum transistors

♦ AFM cantilever arrays, called Millipede, for data storage

♦ Quantum corrals formed using STM to place atoms one by one

♦ AFM and STM tips for nanolithography

♦ Dip-pen nanolithography for "printing" molecules

♦ Nano patterned rigid magnetic disks

♦ Read/write magnetic heads

Summary and Conclusions

➤ Nanoelectronics is the branch of nanotechnology that deals with electronic components and devices.

➤ Some examples of nanoelectronics devices or components are molecular switches, nanotubes, nanowires, and NEMS.

➤ The fabrication techniques to build nanoelectronics devices include lithography vapor deposit methods.

➤ Nanotechnology currently has two active fields of research:

  o The characterization and **fabrication of individual components** that could replace the macroscopic/microscopic silicon components with nanoscale systems.

  o Investigation and **fabrication of potential interconnects**.

➤ In microelectronics, Moore's law has run into a dead end; nanotechnology can help.

➤ Carbon nanotubes can be used as conductors, semiconductors, transistors, and other electronic components and devices.

➤ Nanowires can be used to create n-type and p-type semiconductors, p-n junctions, logical gates, transistors, and field emitters.

➤ Electron microscopy helps to facilitate electronics fabrication and characterization at nanoscale.

Self Test Exercises

1 A 100 micrometer long nanowire has a length to diameter ratio of 30,000. What is the diameter of this nanowire?

2 How many unique numbers can you make with 10 binary bits?

3 A binary bit can represent any of the two digits: 0 or 1. A qubit can represent any of how many digits?

4 A molecular switch is a molecule that can be shifted back and forth between two or more stable states. Give some examples of the variables that can cause shift.

5 The emission of electrons from the surface of a material due to the presence of external electric field is called:

8

Introduction to Nanobiotechnology

Learning Objectives

1. Definitions

2. Unity Behind Diversity of Life

3. Five Fundamental Nanostructures of Life

4. Biotechnology of Proteins

5. DNA Functions and Structure

6. Lipids and Enzymes

7. Bioelectricity

8. Nanobio devices

9. Summary and Conclusions

10. Self Test Exercises

Nanobiotechnology:What Is It?

Biotechnology:

According to the UN Convention on Biological Diversity, biotechnology is any technology or technological application that uses biological systems, living organisms, or derivatives thereof, to make or modify products or processes for specific use.

In other words:

Biotechnology is the study and manipulation of living things and their components to create products and applications.

Nanobiotechnology:

The biotechnology at nano scale ➜

Study existing nanostructures of life to make new structures and to create products and applications.

Three Fundamental Secrets of Nature

- ❖ **Building blocks**. All macroscopic things are composed of smaller building blocks. This building may occur at various levels and layers, for example, all materials around us are made of molecules, molecules are made of atoms, which in turn are made of subatomic particles such as proton, neutron, and electron; and so on.

- ❖ **Design**. Nature does its most important and fundamental design work at smaller scale, that is, at the level of building blocks. This means to really understand a macroscopic structure, you will need to understand its smaller building blocks.

- ❖ **Unity**. There is an underlying unity behind apparent diversity. This is a key point to understand not only phenomena around us but also the diversity of things and phenomena. This unity to some extent can eb traced back to the building blocks.

An Example: Unity Behind Diversity of Life

Diversity of Life:

Number of microorganisms on our planet $\geq 10^{31}$

Number of species of microorganisms $\approx 3 \times 10^9$

Humans on Earth $\approx 6 \times 10^9$

Number of cells in average adult human body $\approx 1 \times 10^{13}$

Unity Behind Diversity:

All living organisms, animals and plants, are made of cells!

All cells are largely made of and run by five fundamental nanostructures!

 *What are those five fundamental nanostructures?*

Five Fundamental Nanostructures of Life

- ◆ All living organisms are made of cells.
- ◆ A cell is largely made of and run by five types of molecules:
    - ○ **Carbohydrates**. The energy molecules
    - ○ **Proteins**. The workforce of life
    - ○ **DNA/RNA**. The information molecules to develop workforce
    - ○ **Lipids**. The infrastructure molecules
    - ○ **Enzymes**. The production machines

 *How are these five types of molecules built?*

How are the Molecules of Life Built?

These molecules are polymers made of smaller building blocks called monomers through chemical reactions that link two monomers together.

For example, in case of carbohydrates:

```
monomer + short polymer ══════▶ longer polymer + water
molecule
```

It's called dehydration reaction.

More complex carbohydrates are built from simpler carbohydrates.

BTW, what are carbohydrates?

Carbohydrates

Commonly called *carbs*.

 A **carbohydrate** is a sugar composed of carbon, hydrogen, and oxygen.

Examples:

- Small sugar molecules dissolved in soft drinks
- Long starch molecules that we consume while eating pasta and potatoes

composition: Made of carbon, hydrogen, and oxygen

Function: Source of energy.

Classification:

- Monosaccharides
- Disaccharides
- Polysaccharides

 An average American consumes 140lbs of sugar per year.

 *What are monosaccharides?*

Monosaccharides

 **Monosaccharides**. Single sugars, the monomers that cells use to make polysaccharides

Examples:

- ◆ **Glucose**. Most important simple sugar; exists in corn, grapes, and blood. Glucose in blood stream supplies energy to all of the body cells.
- ◆ **Galactose**. Less sweet than glucose; found in dairy products.
- ◆ **Fructose**. Molecule that provide fruits and honey their sweetness. Used in corn syrup, ice creams, food and beverages.
- ◆ **Deoxyribose**. Found in the DNA molecule.
- ◆ **Ribose**. Found in the RNA molecule.

Usage:

- ✓ Function as energy molecules.
- ✓ Biotechnologists use glucose as the food source for cell cultures.
- ✓ Cells break glucose to release energy in a form that can be used by the cells; a process called *cellular respiration*.

1. All monosaccharides are converted to glucose during normal metabolic process.
2. Glucose is stored in larger carbohydrates such as disaccharides and polysaccharides.

Disaccharides and Polysaccharides

Disaccharides made of two simple types of sugar molecules: monosaccharides.

Examples:

Sucrose. Molecule in which some plants store glucose. Combination of glucose and fructose. Also called table sugar, cane sugar, beet sugar.

Lactose. Molecule in which mammals store glucose. Gives milk its sweet taste. Combination of glucose and galactose.

 **Polysaccharide**. A polymer of several monosaccharide units.

♦ Complex carbohydrates: molar mass in the range of 4k – 150 M g/mol
♦ Large molecules => Excellent structural and energy storage units.

Examples of energy storage polysaccharides:

✓ Plant starch: amylose
✓ Animal starch: glycogen

Examples of energy structural polysaccharides:

✓ Cellulose: in plant cell walls
✓ Chitin: in fungal cell walls and in insect exoskeletons

Proteins

 **Protein**. One or more polypeptides folded and coiled into a three dimensional structure. A polypeptide is a polymer of many amino acids linked together through bonds called *peptide bonds*.

Greek origin: *proteios* means "of first importance".

- ✓ Makes more than 75% of the dry mass of a cell.
- ✓ Proteins are the work horses of cells.
- ✓ Support wide range of biological functions.
- ✓ A typical cell produces more than 2000 different types of proteins.
- ✓ Human body has tens of thousands of different kinds of proteins; each kind supports a specific functionality.

Composition: A protein is a polymer made of monomers called amino acids selected from a set of 20 types of amino acid molecules called alpha amino acids.

Amino acid. A molecule that contains both:

- An amine group: NH_2
- A carboxylic acid group: $COOH$.

 *What is a peptide bond?*

Peptide Bond

Polypeptide. A protein molecule may consist of 100 or more amino acid molecules grouped together into a chain called polypeptide.

Peptide bond. The bond between two adjacent amino acid molecules in a polypeptide chain: bond that forms between a carboxyl group of one molecule with the amine group of the other molecule by releasing a molecule of water.

 *A protein is not just a polypeptide chain. It is a molecule of unique shape that consists of one or more polypeptide chains twisted, folded, and coiled in a precise fashion to make the shape of the molecule.*

♦ The shape of a protein molecule is sensitive to the environment: temperature, pH, etc.
♦ The change of shape of a protein due to the change in the environment is called *denaturation*.

Examples of Denaturation:

- Boiled egg; the egg white transforms from clear to opaque due to the denaturation of proteins in the egg white.
- Extremely high fevers (> 104°F) denaturate some proteins in the body.

Each kind of protein supports a specific functionality

Functions of Proteins

Recall: Each type of protein corresponds to a specific function.

Some examples:

Functional Group	Examples	Functions
Contractile	1. Actin 2. Myosin 3. Tubulin	1. Muscle contraction 2. Muscle contraction 3. Component of spindle fibers that move chromosomes during cell division
Storage	Found in seeds and eggs	Provide source of amino acids for developing (embryo) plants and animals.
Structural	1. Collagen 2. Fibrin 3. Keratin	1. Component of skin and bones 2. Fibers of a scab 3. Component of hair and nails.
Transport	1. Cytochrome 2. Hemoglobin 3. Lipoprotein	4. Moves electrons through the electron transport system of the cell 5. Carries oxygen in blood 6. Carries cholesterol to bloodstream

Function of a protein type is determined by its structure

Protein Structure

Structure Level	Determined by	Comment	Analogy
Primary	Types of amino acids and their order in the polypeptide chain	Order is coded in the DNA	A metallic wire or a phone cable
Secondary	*H-bonds* Cause stretches on the polypeptide chain	α-helix and β-pleated sheet: the most stable conformations possible	Spiral wire, helix of the curly telephone chord
Tertiary	Attraction or repulsion by charges of the R groups in amino acids; causes additional folding	Provides the overall three dimensional shape	The helix of the telephone chord curled back upon itself
Quaternary	Intermolecular forces holding the two polypeptide chains together.	Applies only for proteins that have more than one polypeptide chain	More than one telephone chords entangled with each other

 *What are H-bonds?*

Hydrogen Bonds

Hydrogen bond. A weak chemical bond that is formed when a slightly positively charged hydrogen atom of a polar molecule is attracted to a slightly negatively charged atom of another polar molecule.

The 3-D structure of a protein is often dependent on a network of H-bonds. These can occur between a variety of atoms, including:

1. Backbone atoms on two different amino acids
2. Backbone atoms and water molecules at the protein surface
3. Atoms on two different sidechains of amino acids
4. Atoms on amino acid sidechains and water molecules at the protein surface
5. Atoms on amino acid sidechains and protein backbone atoms

Any example of a protein structure?

Protein Structure: Hemoglobin

1. First protein with the complete structure discovered.
2. Four polypeptide chains
3. Found in red blood cells that serve to transport oxygen throughout the body

 *Each human red blood cell contains 300 million hemoglobin molecules!*

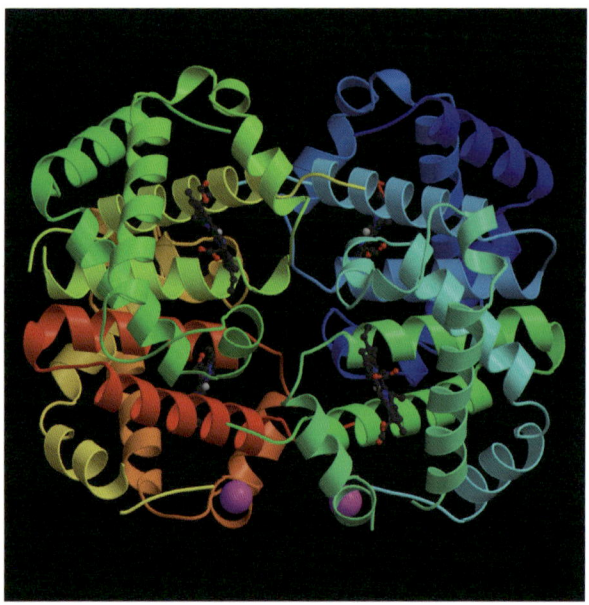

Figure 8.1 One of dozens of structures of hemoglobin in the RCSD protein data bank.

What are the protein products?

Biotechnology: Protein Products

Two types of protein products:

1. **Protein molecules**.

   Example: *Recombinant Insulin*. Regulates blood sugar; used to treat some forms of diabetes.

2. **Protein molecules as a key ingredient**.

   Example:

   Proteins (Enzymes) in contact lens cleaner

 *Proteins in our body cells are made according to the instructions coded dn in the DNA.*

DNA. Deoxyribonucleic acid; a nucleic acid.

Nucleic => exist in the nuclei of eukaryotic cells

Nucleic acid composition:

A polymer, also called polynucleotide, composed of four types of monomers called nucleotides. Each nucleotide has three parts:

Base. A nitrogen containing cyclic molecule

Sugar: A deoxyribose in DNA and a ribose in RNA

Phosphate. One or more phosphate groups

 *What are the functions of DNA?*

DNA Functions and Structure

Store information that has directions (or instructions) for

♦ Building RNAs
♦ Building proteins
♦ Regulating the use of genetic information

Bonds that make DNA structure:

What binds the nucleotides together in a polynucleotide?

Nucleotides are joined together by covalent bonds between the sugar of one nucleotide and the phosphate of the other.

What binds the two strands of the DNA together?

Weak bonds between bases called hydrogen bonds or H-bonds.

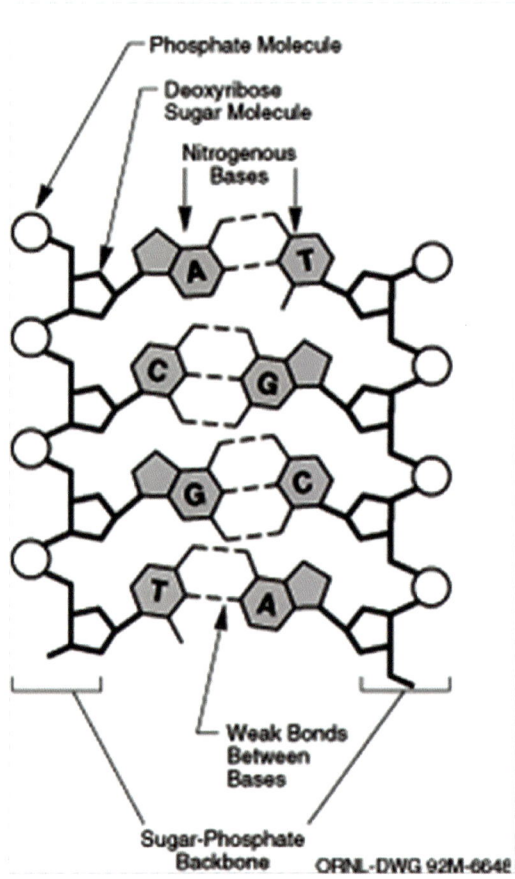

Figures 8.2 Two strands of DNA bonded together through H-bonds.

DNA Structure

Two polynucleotides wrapped around each other to form a double helix:

- Double helix with a diameter of about *2 nm*.
- Two DNA strands held together by *H-bonds* between bases: A bonds (pairs) only with T and vice versa, and C bonds only with G and vice versa.
- A strand is a polymer of nucleotides with a specific sequence of four types of bases: A, C, G, and T. This sequence determines the code (genetic information).
- The two ends of a strand are marked as 5` and 3` owing to the positions of a carbon atom of interest in the sugar molecules at the ends.

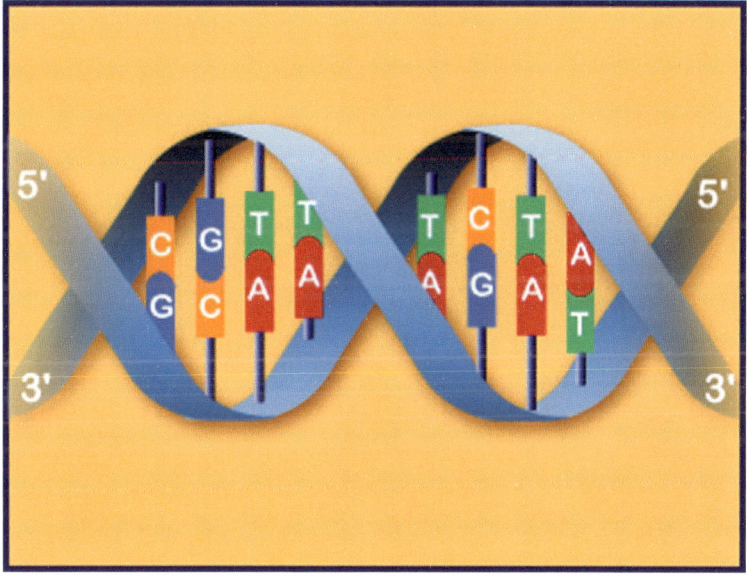

Figure 8.3 Illustration of a DNA helix.

 *A typical cell in your body has about two meters of DNA => needs packaging*

DNA Packaging

Chromosome. A structure into which a DNA is organized; it also contains some proteins which serve to package the DNA and control its functions.

Genome. A complete set of DNA for an organism: contains the whole hereditary (genetic) information and the non coding sequences.

♦ Each human cell contains a complete genome.

♦ Human genome consists of 24 distinctive chromosomes: Each chromosome is a physically separate molecule.

♦ A human cell consists of 46 chromosomes: 22 homologous pairs + 2 sex chromosomes.

♦ Chromosomes range in length from 50 million to 250 million base pairs: 3 billion/24 = 125 million

Gene. A basic physical and functional unit of heredity. Genes comprise only about 2% of human genome. Rest: regulating functions + unknown (Junk DNA)

Regulating example: where, when, and what quantity of protein to make.

The human genome consists of about 20,000 to 25,000 genes

The smallest known genome is that of a bacterium and contains 600,000 base pairs (A-T, C-G). Human genome consists of about 3 billion base pairs.

Cell: A Protein manufacturing Factory

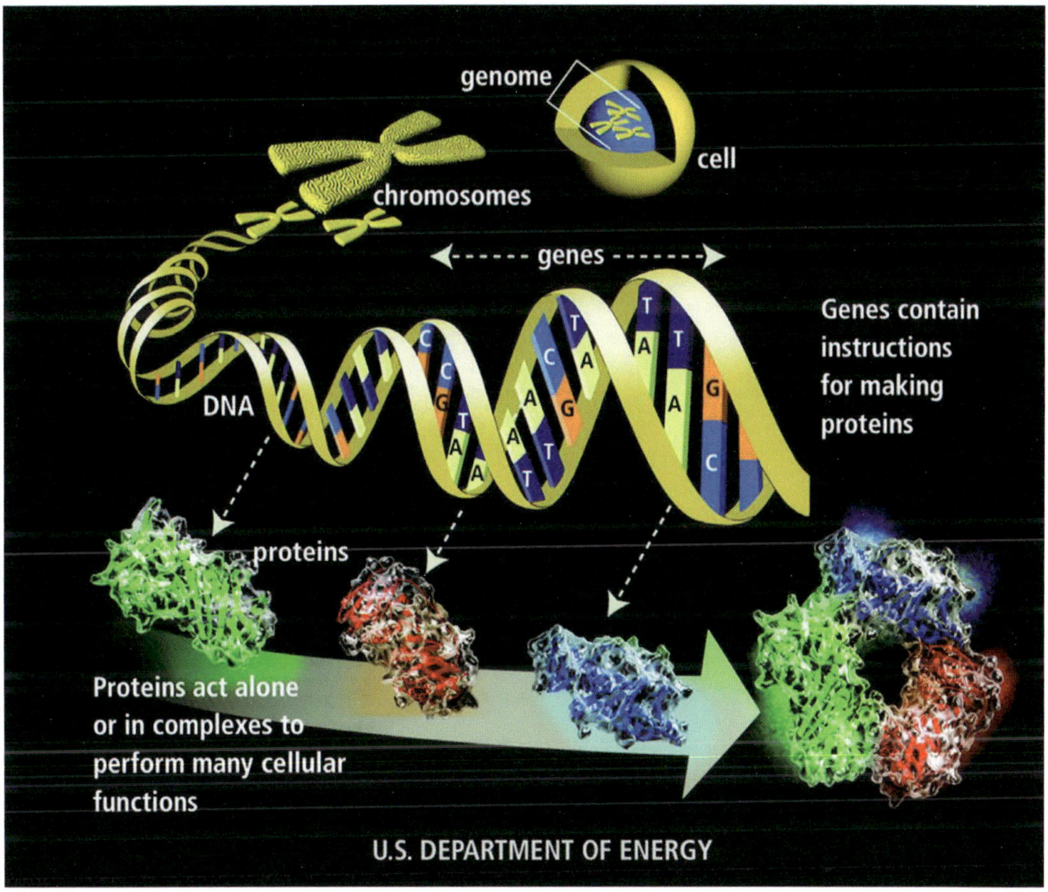

Figure 8.4. Illustration of what happens inside a cell to produce proteins.

How is DNA used in biotechnology?

Biotechnology of DNA

- DNA synthesis
- Cloning
- DNA sequencing
- Discovering genetic origin of a disease, e.g., monitoring and comparing the activity of tens of thousands of genes simultaneously in cancerous and noncancerous tissues.

Another important nanostructure of life is lipid.

 *What are lipids?*

Lipids

 **Lipids**. A family of biomolecules that have the common property of being soluble in organic solvents but insoluble in water.

Basic characteristics:

✓ Largely built from carbon and hydrogen atoms; very few oxygen atoms if any.

✓ Used as long term storage of energy: Oil (plants)…fats (animals).

✓ Dual-nature structure of lipid molecule:

  o A polar head that is *hydrophilic*, means water soluble

  o A hydrocarbon tail that is *hydrophobic*, means insoluble in water

✓ Basic building blocks to build infrastructure:
  - Storage of energy in fats.
  - Cholesterols
  - Sex hormones and gender traits
  - Membranes, e.g. around the cells and organelles in the cells: key components cholesterols and phospholipids

A special type of proteins is called enzymes →

Enzymes

 **Enzymes**. Type of proteins that catalyze biological reactions.

- ◆ **A type of proteins**, usually listed separately due to its importance.
- ◆ **A biological catalyst, i.e.,** changes the reaction rate without changing itself in the process.
- ◆ Lowers the **activation energy** required by the reactants to start a chemical reaction. Enzymes do this by binding to reactant molecules and putting on them a physical or chemical stress.
- ◆ Make **metabolism** work at cooler temperatures.

 *An enzyme is a bionanomachine that is designed for a specific reaction to produce a specific product. It typically increases the reaction rate (production rate) trillion fold.*

 *Where is nano involved in all these molecules of life?*

Typical sizes of various biological particles/structures

Particle	Molecular Mass (g/mole)	Size: d (nm)	Type	Comment
Glycine	75	0.42	Amino acid	Simplest of the 20 standard amino acids.
Tryptophan	204	0.67	Amino acid	One of the 20 standard amino acids. Essential in human neutrition.
Cytosine monophosphate	309	0.81	Nucleotide	Smallest DNA nucleotide
Guanine monophosphate	361	0.86	Nucleotide	A DNA nucleotide on the larger end of size spectrum
Insulin	5,800	2.2	Protein	A polypeptide hormone that regulates carbohydrate metabolism
Hemoglobin	68,000	7.0	Protein	Carries oxygen in human blood
Lipoprotein	1, 300,000	20	Lipid and protein	Carry fats around the body
Fibrin	400,000	50	Protein	Involved in clotting of blood

Possibility of some interesting applications ➔

Bioelectricity with the Help of Nano

Bio solar Cell Inside of a Chloroplast

Possibility of drawing electricity directly from biological cells

How?

An Example:

1. Use the Algae (unicellular) cell
2. Use the solar energy to split water into oxygen, protons, and electrons driven by the excitation of the two photosystems of the photosynthetic electron transport system, located in the thylakoid membranes of chloroplasts.
3. Use nano-probe systems to locally capture the generated high energy electrons in order to produce electricity.

Production of High Energy Electrons:

$$2H_2O + \text{Solar Energy} \longrightarrow 4H^+ + O_2 + 4e^-$$

 *What does the bio solar cell look like?*

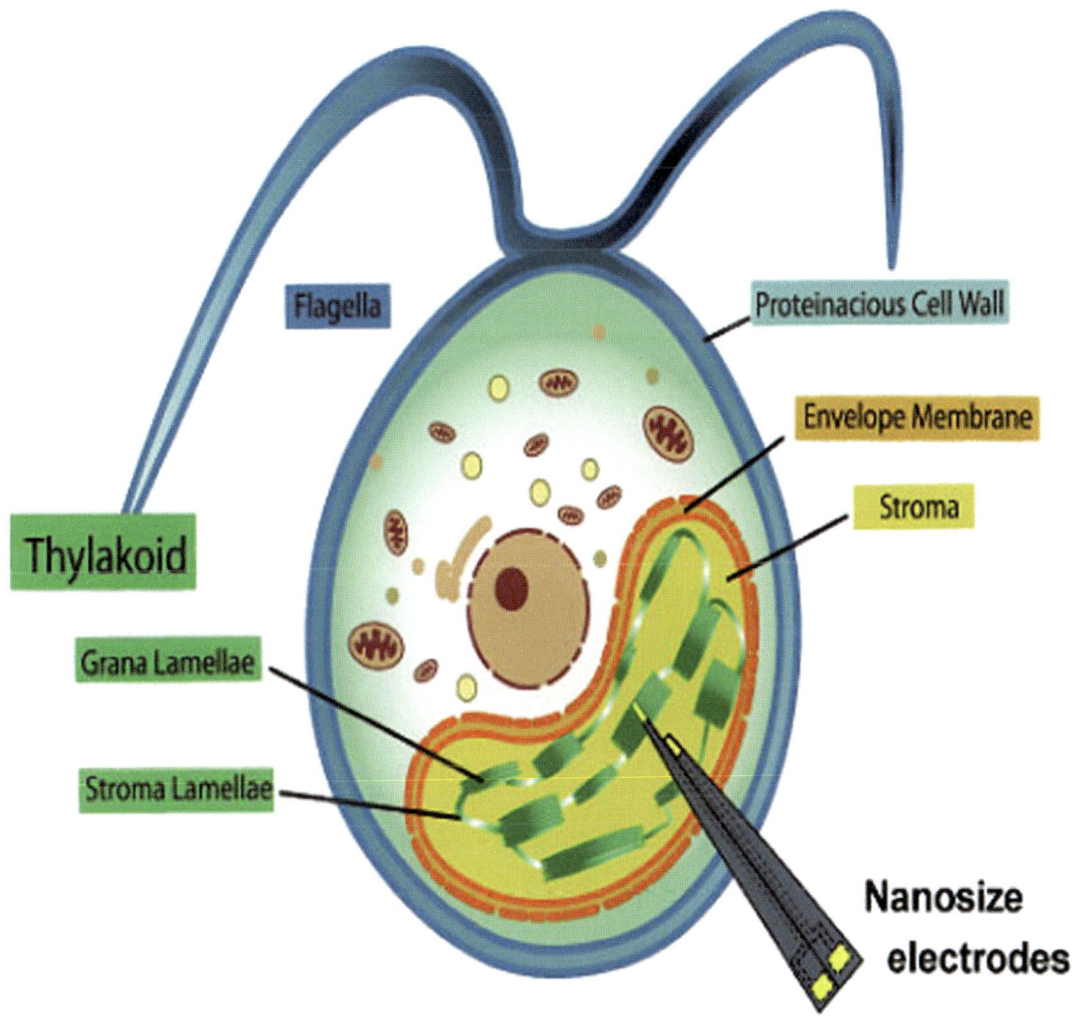

Figure 8.5 Schematic of bio solar cell using two nano-scale electrodes.

Developing Nanobio Devices

You develop a nanobio device by using what is called structural DNA nanotechnology.

What is structural DNA nanotechnology?

It refers to the construction of nano-sized molecules from the basic DNA components as the building blocks. The construction process makes use of a technique called RFLP (Restriction fragment length polymorphism). **Here is the basic process:**

1. DNA is placed with a restriction enzyme.
2. The restriction enzyme cuts the DNA into fragments.
3. The DNA fragments are separated and collected by using an appropriate technique such as electrophoresis.
4. Using their polarity and topological structures, you make these DNA fragments (or strands) interact with each other to create a progressively complex structure. For example, two DNA strands can interact with each other to make a double helix, and this double helix can interact with another strand to make a more complex structure, and so on. This interaction is called *reciprocal exchange.*
5. The reciprocal exchange leads to successively more and more complex structure until you get the desired structure.

Can you please give some examples of end products of reciprocal exchange?

Structures from DNA

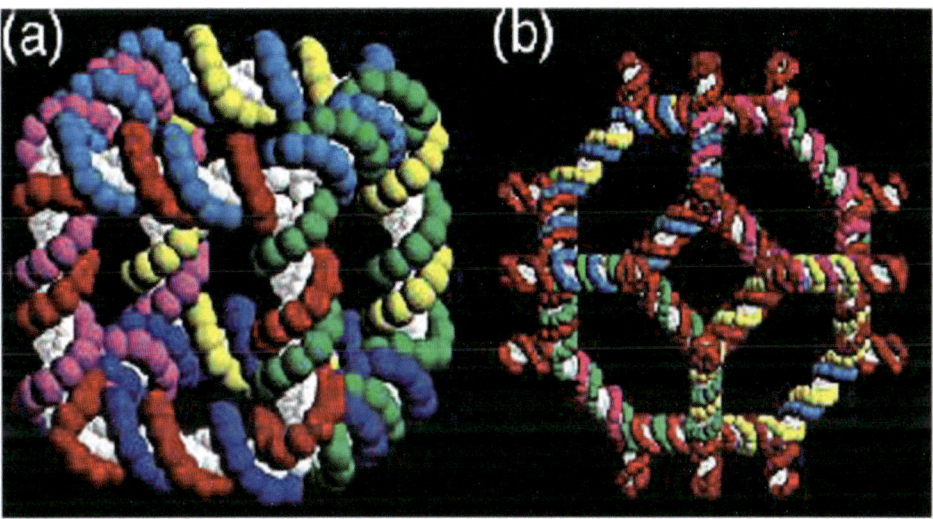

Figure 8.6 Examples of DNA structures developed by reciprocal exchange.

a) Cube like structure of DNA

b) Double helical strands combined to form the square faces and the hexagonal faces of a structure

These structures act as nanomachines or nanomotors to perform various tasks such as produce a desired protein. You can also call it a nanobiodevice.

 *OK, you have developed a nanobio device, inserted it into an organism, but how will you power it?*

Powering a Nanobio Device

A very good candidate that can potentially be used as a power supplier to the nanobio devices inserted into the body is the natural biological pump that exists inside mitochondria and is called *F1F0- ATPase*, an enzyme.

ATP. Adenosine triphosphate; the common currency of chemical energy in most cells.

The main purpose of metabolizing food is to draw energy from it to make ATP. This energy is then used to drive all cellular processes needed to survive and reproduce.

F1F0-ATPase drive the synthesis of ATP in a cell:

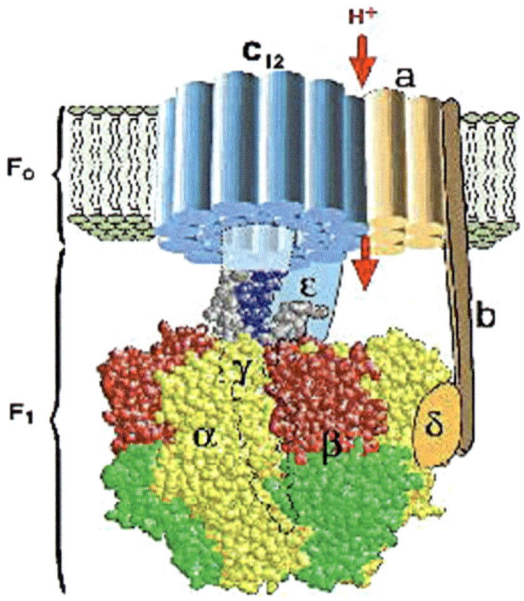

Figure 8.7 F1 ATPase acting as a biological pump that exists in mitochondria.

Here is how it works. A proton moves down the concentration gradient, which gives the enzyme a spinning motion. This unique spinning motion bonds adenosine diphosphate (ADP) and Pi (inorganic phosphate) together to form ATP.

ATPase as a nanomotor

Another look at F-ATPase:

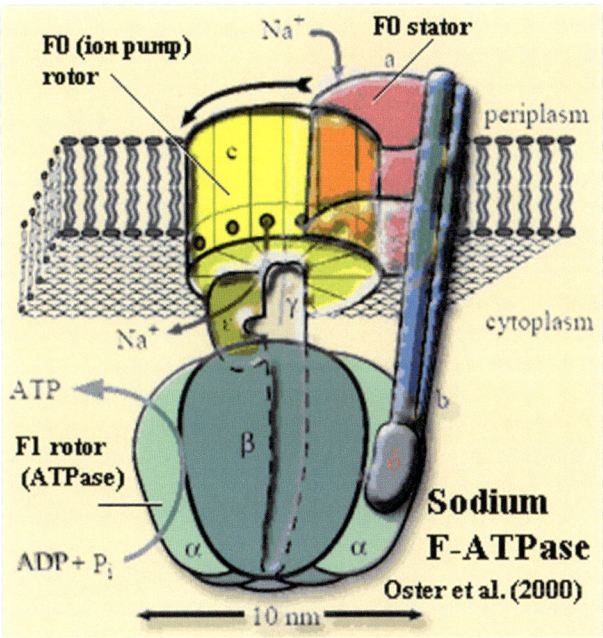

Figure 8-8. Illustration of F-ATPase found in mitochondrial inner membranes and in chloroplast thylakoid membranes.

❖ **Consists of two domains:**

> **F0.** Integral part of the membrane.

> **F1.** Peripheral; at the side of the membrane to which the protons are moving.

❖ Each of both domains is assembled of units called polypeptides.
❖ These units together form a rotary motor.

❖ As the protons bind to the units of domain F0, they make parts of it to rotate.
❖ The rotation is propagated to F1 domain.
❖ ADP and inorganic phosphate (Pi) bind to three units of F1 called beta (β).
❖ Every time it goes through 120° rotation an ATP is released.

Bottom Line:

FTPase uses a proton gradient to drive ATP synthesis:

1. Allowing the flux of protons across the membrane down their electrochemical gradient.
2. Use the energy released by the transport reaction to synthesise ATP from ADP and inorganic phosphate.

In some bacteria, sodium ions may be used instead of hydrogen ions (protons).

 OK, you have a nanobio device, and a nanomotor to supply power to it, what will you do with this device?

Uses of Nanobio devices

♦ Disperse the devices through some liquid medium.

♦ Get them into the organism (body) by using an appropriate method such as swallow, inject, or implant.

♦ Power them.

♦ Let them do the job they are programmed to do:

- Recognize problematic cell.
- Cure or destroy the problematic cell.
- Deliver drug to a specific target in the body of an organism.

Challenge. *The nanobio devices injected into the body of an organism has the risk of being rejected as antigens.*

Summary and Conclusions

➢ Life is supported by nanostructures such as molecules like carbohydrates, DNA, lipids, and proteins.

➢ Nanobiotechnology involves studying existing nanostructures of life to make new structures and to create products and applications.

➢ Protein molecules are either products themselves or ingredients for other products in biotechnology.

➢ DNA can be used to create proteins because it contains instructions to create proteins.

➢ Enzymes act as catalysts for biological reactions.

➢ Lipids are used to build membranes, e.g. around the cells and organelles in the cells.

➢ There is a possibility of drawing electricity from biological cells.

➢ Nanobio devices can be built by using structural DNA technology.

➢ Nanobio devices can be used to detect and cure or destroy problematic cells and to deliver drugs to specific targets in an organism.

Self Test Exercises

1 Name the nanostructures that function as workhorses of cells.

2 Energy in our body is stored in which nanostructures?

3 These molecules make more than 75% of the dry mass of a cell:

4 The bond between two adjacent amino acid molecules in a polypeptide chain of a protein molecule is called:

5 The diameter of a DNA double helix is about:

6 The bonds that keep two strands of a DNA helix together are called:

7 This protein (enzyme) drives the synthesis of ATP, an energy molecule in a cell:

8 What is the main challenge in using nanobio devices in a human body?

9

Environmental, Health, and Safety Implications of Nanotechnology

Learning Objectives

1. Definitions
2. Classification of Pollutants
3. The Origin of EHS Issues
4. Properties of Nanoparticles Relevant to EHS Issues

5. *Risk Assessment*

6. *Exposure*

7. *Hazard*

8. *Toxicity*

9. *Risk Characterization of EHS*

10. *Laws and Regulations*

11. *EHS Research Needs*

12. *Summary and Conclusions*

13. *Self Test Exercises*

Definitions

- ♦ **Environmental Pollution.** Introduction of contaminants into the environment.
- ♦ **Pollutant.** The contaminant that pollutes.
- ♦ **Hazard.** A situation or an object that poses risk to environment, health, life, or property.
- ♦ **Toxin.** A poisonous substance produced by living organisms or cells that is active even at very low concentrations. It could be in form of molecules such as proteins.

What kinds of pollutants are out there?

Classification of Pollutants

Main Classes of Nano-emissions:

- **Particulate.** Tiny particles of solid or liquid suspended in a gas.

- **Gas.** A state of matter that consists of a collection of particles without a definite shape and the particles are in random motion.

- **Aerosol.** Mix of particles with a specific gas or gases.

Classification of Pollutants Based on What They Pollute:

- Air Pollutants

- Water Pollutants

- Land Pollutants

 *Toxins are capable of causing diseases when enter into body tissues and also causing damage by neutralizing antibodies and antitoxins.*

Please download the complete chapter by using the following link:

http://www.infonentialinc.com/nanotech/nanoehs.pdf

Answers to Self Test Exercises

Module 1

1. **True or False**: Nanoscale is larger than micro scale but smaller than macro scale.

Solution:

False. Nanoscale that is typically considered in the range of 1 nm (10^{-9} m) to 100 nm, is part of what physicists referred to as micro scale as opposed to macro scale, and it goes all the way from micro (10^{-6}) to as small as possible.

2. Write the following quantities in powers of 10:

75 nanometer

500 nanogram

35 millimeter

Solution:

A.

75 nanometer = 75 x10^{-9} m = 7.5 x x10^{-8} m

B.

500 nanogram = 500 x10^{-9} g = 5x10^{-7} g

C.

$m = 35 \times 10^{-3} m = 3.5 \times \times 10^{-2} m$

3. Convert the following:

0.25 g = _____ μg

3.5 μL = _____ nL

560 mm = _____ cm

2.5×10^{8} μg = _____ kg

0.055 m = _____ mm

136 mL = _____ L

Solution:

A. 0.25 g = _____ 2.5×10^{5} ___ μg

$0.25 g = 0.25 (\mu/10^{-6}) g = 2.5 \times 10^{5}$ μg

B. 3.5 μL = ___ 3.5×10^{3} ___ nL

3.5 μL $= 3.5 \times 10^{-6} L = 3.5 \times 10^{-6} \times (n/10^{-9})L = 3.5 \times 10^{3}$ nL

C. 560 mm = ___ 56 ___ cm

D. 2.5×10^{8} μg = ___ 0.25 ___ kg

2.5×10^{8} μg $= 2.5 \times 10^{2} g = 2.5 \times 10^{2} \times (k/10^{3}) g = 2.5 \times 10^{-1}$ kg $= 0.25$ kg

E. 0.055 m = ___ 55 ___ mm

F. 136 mL = ___ 0.136 ___ L

4. Name some fields that merge at the nano junction .

Solution:

Physics

Chemistry

Biology

Material science

Engineering

Computer science

5. Give examples of two structures that are studied in wet nanotechnology.

Solution:

DNA

Proteins

6. Give examples of two structures that are studied in dry nanotechnology.

Solution:

Nanotube

Buckyball

7. Name two nano tools.

Solution:

Electron microscope

Atomic force microscope

8. The material world at nanoscale is governed by what branch of physics?

Solution:

Quantum mechanics

Module 2

1. Explore nanotech positions at the following websites:

http://www.workingin-nanotechnology.com/

Solution:

Do it yourself

2. Asia is matching dollar with dollar on government funding in nanotechnology. What does it tell you about nanotechnology?

Solution:

It's evidence that nanotechnology has gained world-wide credibility. It also means that the US government has so far underestimated the importance of nanotechnology partially due to its extremes stance on issues such as stem cell research.

3. What is NNI?

Solution:

National Nanotechnology Initiative: A US Government program to unify the nanotechnology efforts by the government at one platform.

4. What is NIH?

Solution:

National Institute of Health, a U.S. government agency.

Module 3

1. Name some first generation nano products.

Solution:

- Bowling balls that are harder
- Golf balls that fly straighter
- Nano car wax that gives you a shinier looking vehicle because it fills in tiny cracks more effectively than the standard wax
- Smart clothing:
  - ✓ Pants that repel wax
  - ✓ Shirts that don't stain
  - ✓ Socks that don't stink due to the inclusion of nano-sized silver particles
- Tennis balls that last longer
- Tennis rackets that are stronger

2. Nanotechnology has something to do with the evolution of video games from arcade size of the past such as Pong, Frogger, and PacMan to the modern more sophisticated games being played in homes on platforms such as?

Solutions:

♦ X Box

♦ Play station

♦ Game Cube.

3. Name the types of nanotechnology product types in spatial dimensions.

Solution:

1. Nano

2. Micro

3. Macro

4. These nanoparticles are used in sun creams because they absorb ultra violet (UV) light:

Solution:

Nano-sized zinc oxide particles

5. Proteins and nucleic acids are used as probe molecules in what kind of sensor applications?

Solution:

Biosensor applications

Module 4

1. Why can't you use the optical microscope to study nanostructures? (Hint: resolution and wavelength)

Solution:

The minimum wavelength of visible light is about 400 nm, which is not small enough to probe with a resolution better than 100 nm, that is, less than 100 nm.

2. Rearrange the entries in the second column of the following table to correct errors:

Technique	Feature
Microscopy	Used for analysis
Spectroscopy	Used for visualization
Optical microscope	Uses Electromagnetic force
Electron microscope	Uses electromagnetic waves
Atomic force microscope	Uses electron waves

Solution:

Technique	Feature
Microscopy	Used for visualization
Spectroscopy	Used for analysis
Optical microscope	Uses electromagnetic waves (light)
Electron microscope	Uses electron waves
Atomic force microscope	Uses Electromagnetic force

3. Rearrange the entries in the second column of the following table to match them correctly with the entries in the first column.

Type of Spectroscopy	Basic Operational principle
IR	Photons in, electrons out
X-ray diffraction	Photons in, phonons out
X-ray photoelectron emission	Photons in
Raman	Photons in, photons out

Solution:

Type of Spectroscopy	Basic Operational principle
IR	Photons in
X-ray diffraction	Photons in, photons out
X-ray photoelectron emission	Photons in, electrons out
Raman	Photons in, phonons out

4. Rearrange the entries in the second column of the following table to match them correctly with the entries in the first column.

Type of Microscope	Basic Operational principle
TEM	Primary electrons in, secondary electrons out
SEM	Electrons in, electrons out.
STM	The current between the scanning probe and the specimen surface is monitored
AFM	The atomic force between the scanning tip and the specimen surface is monitored

Solution:

Type of Microscope	Basic Operational principle
TEM	Electrons in, electrons out.
SEM	Primary electrons in, secondary electrons out
STM	The current between the scanning probe and the specimen surface is monitored
AFM	The atomic force between the scanning tip and the specimen surface is monitored

5. Which microscope uses the principle of quantum mechanical tunneling?

Solution:

Scanning tunneling microscope (STM)

Module 5

1. Calculate the energy of a proton at rest.

Solution:

The effective mass of a proton at rest:

$m = m_0 = 9.11 \times 10^{-31}$ kg

$E = mc^2$

$= 1.673 \times 10^{-27}$ kg $\times 9.00 \times 10^{16}$ m^2/s^2

$= 15.057 \times 10^{-11}$ kg- m^2/s^2

$= 15.057 \times 10^{-11}$ J

$= 15.1 \times 10^{-11}$ J $\times \dfrac{1\,ev}{1.60 \times 10^{-19}\,J}$

$= 9.44 \times 10^8$ ev

$= 944$ MeV

2. What is the name of the electromagnetic force that holds buckyballs together?

Solution:

van der Waals force

3. The Young's modulus Y for carbon nanotubes is in the range of 1.3 to 1.8 TPa, that is, about 10 times that of steel. Does that mean that carbon nanotubes are more flexible than steel?

Solution:

No, it means carbon nanotubes are less flexible. The higher the value of Y, the less flexible the material is.

4. A single walled carbon nanotube has a diameter of `1nm` and the length to diameter ratio of `10,000`. What is the length of this nanotube in micrometers?

Solution:

```
Length/diameter = 10000
Diameter = 1 nm
```

```
Length = 10000xdiameter = 10000x1nm = 10000nm = 10 μm.
```

5. A nano planar material has a lot of energy at the edges. Left alone, it will automatically roll itself into what to minimize the energy?

Solution:

The most stable state is the state with minimum energy. So the nano planar material will roll itself into a ball or a tube to minimize the energy by getting rid of the energy at the edges.

Module 6

1. Matter confined to a nanoscale makes what kind of particles?

Solution:

Nanoparticles

2. The electrical properties of semiconductors can be controlled by what characteristic of semiconductors?

Solution:

Band gap

3. In a quantum well, particle is free to move in how many spatial dimensions

Solution:

The particle in a quantum well is free to move in two dimensions because quantum well offers confinement in one dimension.

4. The optical and electronic properties of quantum confined structures exhibit a strong dependence on the dielectric constant of the surrounding media. Why is that important?

Solution:

We can control the properties of the nanostructures by tuning the dielectric constant of the surrounding media.

5. Rearrange the entries in the second column of the following table to match them correctly with the entries in the first column.

Entity	Definition
Plasmon	Quanta of light
Photon	Quanta of oscillations in plasma
Plasma	Ionized bulk matter that contains free electrons
Plasma waves	Changing density of electric charges at a given point

Solution:

Entity	Definition
Plasmon	Quanta of oscillations in plasma
Photon	Quanta of light
Plasma	Ionized bulk matter that contains free electrons
Plasma waves	Changing density of electric charges at a given point

Module 7

1. A 100 micrometer long nanowire has a length to diameter ratio of 30,000. What is the diameter of this nanowire?

Solution:

L/d = 30,000

➔ d = L/30,000 = 100   m/30,000 = 3.33 nm

2. How many unique numbers can you make with 10 binary bits?

Solution:

Unique numbers with 10 binary bits = 210 = 1024

3. A binary bit can represent any of the two digits: 0 or 1. A qubit can represent any of how many digits?

Solution:

Infinite

4. A molecular switch is a molecule that can be shifted back and forth between two or more stable states. Give some examples of the variables that can cause shift.

Solution:

Current

pH

Temperature

Light

5. The emission of electrons from the surface of a material due to the presence of external electric field is called:

Solution:

Field emission

Module 8

1. Name the nanostructures that function as workhorses of cells.

Solution:

Proteins

2. Energy in our body is stored in which nanostructures?

Solution:

Carbohydrates

3. These molecules make more than 75% of the dry mass of a cell:

Solution:

Proteins

4. The bond between two adjacent amino acid molecules in a polypeptide chain of a protein molecule is called:

Solution:

Peptide bond

5. The diameter of a DNA double helix is about:

Solution:

2 nm

6. The bonds that keep two strands of a DNA helix together are called:

Solution:

Hydrogen bonds

7. This protein (enzyme) drives the synthesis of ATP, an energy molecule in a cell:

Solution:

F1F0-ATPase

8. What is the main challenge in using nanobio devices in a human body?

Solution:

The nanobio devices injected into the body of an organism has the risk of being rejected as antigens.

Module 9

1. Search the Web to find some examples of VOC.

Solution:

Examples of volatile organic compounds (VOC):

A wide range of carbon-based molecules, such as aldehydes, ketones, and other light hydrocarbons.

The most common VOC is methane.

2. While talking about EHS, why does the size of the particles matter?

Solution:

Onc major reason for the size to matter is the surface area. Most chemical reactions involving solids happen at the surface where chemical bonds are incomplete. For example, chemical reactions happen by breaking old bonds and making new bonds. In nanomaterial more surface area is available.

3. In this exercise, you will realize how the surface area of a given amount of material increases as the size of its constituents decreases.

A. Calculate the surface area of a cube with each dimension equal to 1 cm.

B. Now consider that this cube has been converted into nanopowder that consists of cubes with each dimension equal to 1 nm. Calculate the total surface are of all the cubes.

Solution A:

A cube has six faces.

Area of one face = $1 \, cm^2$.

Total area of the macrocube = $6 \, cm^2$.

Solution B:

Volume of the macrocube = $1 \, cm^3$.

Volume of one nanocube = $1 \, nm^3$.

Number of nanocubes = $1 \, cm^3 / 1 \, nm^3 = 10^{-6}/10^{-27} = 10^{21}$.

Surface area of one nanocube = $6 \, nm^2$.

Total surface area of nanopowder = $6 \times 10^{21} \, nm^2$
$= 6000 \, m^2 = 60 \, million \, cm^2$

Compare this to the surface area of the macrocube, $6 cm^2$.

4. Perform the web search to find out what is in-vitro and in-vivo are.

Solution:

In-vitro. the techniques of performing experiments in a controlled environment outside of a living organism; for example in a test tube.

In-vivo; means it takes place inside an organism. Refers to experimentation performed in or on the living tissue of a whole, living organism as opposed to a partial or dead one or in a controlled environment.

5. Search the Web to find out what *abiotic* effects are.

Solution:

Abiotic means not associated with or derived from living organisms. Abiotic factors in an environment include such items as sunlight, temperature, wind patterns, and precipitation.

Art Credits and Acknowledgments

Unless otherwise acknowledged, all pictures and illustrations in this book are the property of Infonential, Inc. We have made our best effort to trace and acknowledge the ownership of the following items. In the event of any question or issue arising from the use of any of these items, we will be pleased to make the necessary corrections in the future printings.

- Richard Feynman, page 13, particlezoo.wordpress.com

- Page 19, www.cbse.uab.edu/ribbons/help/dna_rgb.html and soe.ucdavis.edu/ss0708/taylork/

- Figure 1.3, Department of Energy: http://www.er.doe.gov/production/bes/scale_of_things.html

- Figure 1.4 and Figure 1.5, NNIN Education Site

- Figure 3.1 news.bbc.co.uk

- Figure 4.3, Laboratoire IMPMC E. Larquet et N. Boisset., CNRS-IMPMC, Paris, 2005; picture in the public domain.

- Figure 4.4, nucleosinc.com

- Figure 4.5, and Figure 4.6, IBM research laboratory, Almaden, California.

- Figure 5.2, mrsec.wisc.edu/Edetc/nanoquest/carbon/

- Figure 5.3, http://staff.science.uva.nl/~rhd/math.html

- Figure 5.4 R. Bruce Weisman, Ph.D., Rice University

- Figure 6.4 Research Institute of Sustainable Energy: www.rise.org.au

- Figure 7-1, Intel.

- Figure 7-2, Zattl Research Group, University of California, Berkeley.

- Figure 8-1, RCSB Protein Data Bank.

- Figures 8.2, Genome Management Information System, Oak Ridge National Laboratory
- Figure 8.3, picture in the public domain: http://members.cox.net/amgough/Fanconi-genetics-genetics-primer.htm
- Figure 8.4, Genome Management Information System, Oak Ridge National Laboratory
- Figure 8.5, www-rpl.stanford.edu
- Figure 8.6, www.nue.clt.binghamton.edu
- Figure 8-7, www.nue.clt.binghamton.edu
- Figure 8.8, www.palaeos.com

References and Further Reading

1. Springer handbook of nanotechnology by Bharat Bhushan (editor), Springer, 2006.
2. www.nano.gov
3. www.fda.gov/nanotechnology
4. www.nsti.org
5. www.nanotech-now.com
6. www.foresight.org

CPSIA information can be obtained
at www.ICGtesting.com
229004LV00004B